“This book offers a unique and thorough exami
be used effectively to deal with the climate crisis
political freedom, doing so through a debate with
of political philosophy. It is required reading for anyone interested in both the political potentials and risks of AI in the emerging age of the Anthropocene.”

— **Pieter Lemmens,** *Radboud University Nijmegen, The Netherlands*

Green Leviathan or the Poetics of Political Liberty

This book discusses the problem of freedom and the limits of liberalism considering the challenges of governing climate change and artificial intelligence (AI). It mobilizes resources from political philosophy to make an original argument about the future of technology and the environment.

Can artificial intelligence save the planet? And does that mean we will have to give up our political freedom? Stretching the meaning of freedom but steering away from authoritarian options, this book proposes that, next to using other principles such as justice and equality and taking collective action and cooperating at a global level, we adopt a positive and relational conception of freedom that creates better conditions for human and non-human flourishing. In contrast to easy libertarianism and arrogant techno-solutionism, this offers a less symptomatic treatment of the global crises we face and gives technologies such as AI a role in the gathering of a new, more inclusive political collective and the ongoing participative making of new common worlds.

Written in a clear and accessible style, *Green Leviathan or the Poetics of Political Liberty* will appeal to researchers and students working in political philosophy, environmental philosophy, and the philosophy of technology.

Mark Coeckelbergh is Professor of Philosophy of Media and Technology at the Philosophy Department of the University of Vienna and former President of the Society for Philosophy & Technology. He has authored 14 books and numerous articles and is a member of advisory bodies such as the European Commission's High-Level Expert Group on AI.

Routledge Studies in Contemporary Philosophy

Extimate Technology
Self-Formation in a Technological World
Ciano Aydin

Modes of Truth
The Unified Approach to Truth, Modality, and Paradox
Edited by Carlo Nicolai and Johannes Stern

Practices of Reason
Fusing the Inferentialist and Scientific Image
Ladislav Koreň

Social Trust
Edited by Kevin Vallier and Michael Weber

Green Leviathan or the Poetics of Political Liberty
Navigating Freedom in the Age of Climate Change and Artificial Intelligence
Mark Coeckelbergh

The Social Institution of Discursive Norms
Historical, Naturalistic, and Pragmatic Perspectives
Edited by Leo Townsend, Preston Stovall, and Hans Bernard Schmid

Epistemic Uses of Imagination
Edited by Christopher Badura and Amy Kind

Political Philosophy from an Intercultural Perspective
Power Relations in a Global World
Edited by Blanca Boteva-Richter, Sarhan Dhouib, and James Garrison

For more information about this series, please visit: https://www.routledge.com/Routledge-Studies-in-Contemporary-Philosophy/book-series/SE0720

Green Leviathan or the Poetics of Political Liberty

Navigating Freedom in the Age of Climate Change and Artificial Intelligence

Mark Coeckelbergh

NEW YORK AND LONDON

First published 2021
by Routledge
605 Third Avenue, New York, NY 10158

and by Routledge
2 Park Square, Milton Park, Abingdon, Oxon, OX14 4RN

Routledge is an imprint of the Taylor & Francis Group, an informa business

Library of Congress Cataloging-in-Publication Data
Names: Coeckelbergh, Mark, author.
Title: Green Leviathan or the poetics of political liberty : navigating freedom in the age of climate change and artificial intelligence / Mark Coeckelbergh.
Description: New York, NY : Routledge, an imprint of Taylor & Francis Group, 2021. |
Series: Routledge studies in contemporary philosophy | Includes bibliographical references and index.
Identifiers: LCCN 2020058243 | ISBN 9780367745998 (hbk) | ISBN 9780367747794 (pbk) | ISBN 9781003159490 (ebk)
Subjects: LCSH: Environmentalism. | Climate change— Social aspects. | Liberty. | Freedom.
Classification: LCC GE195 .C63 2021 | DDC 304.2/8—dc23
LC record available at https://lccn.loc.gov/2020058243

ISBN: 978-0-367-74599-8 (hbk)
ISBN: 978-0-367-74779-4 (pbk)
ISBN: 978-1-003-15949-0 (ebk)

Typeset in Sabon
by codeMantra

Contents

Acknowledgments

I would like to thank the editors from Routledge, Andrew Weckenmann and Allie Simmons, for their enthusiasm and excellent work on this book project. I warmly thank Zachary Storms for organizational assistance during the final stages of preparing the manuscript. This significantly contributed to the timely submission. I would also like to thank Alan Marshall for permission to use his artwork. And, last but not the least, I wish to thank my family and friends for their support and company during this challenging year 2020.

Figure 1 Artwork Tokyo 2121 (Alan Marshall).

1 A Green Brave New World

Dealing with Climate Change and AI as a Political Problem Concerning Freedom

A Very Short Story: 'A Green Brave New World', or 'The Green Panopticon'

For the first time in 20 years, he could see the city again. The warm but mild spring sun and light breeze felt pleasant on his face and he could hear the birds singing. But not just a few. A lot. There were even butterflies again. Many had beautiful colors. He could not even recognize all the species of animals he saw. And so much green everywhere! Trees, plants, actually it hardly looked like a city at all. Not the city he knew. Not the city the green movement has been complaining about since the 1960s. This spring was not silent but noisy and exuberant, full of life and energy. It soon dawned on him that this was no longer the world he knew. Not the civilization he rebelled against. Not the society he wanted to change. This was looking much like the new, better world that he and his ecoterrorist group fought for 20 years ago when they managed to blow up an oil tanker. It seemed that eventually the rebels won. Climate change was controlled. Mass extinction of species was prevented. Life has won. The ecosystems of the planet recovered and even improved. A slow, hopeful smile appeared. As he briefly closed his eyes and breathed in the clean air deeply, he felt the joy of a long-awaited success. Everything had not been in vain. There was a future now. A bright future and a green future. Not only a future for him, leaving the prison after such a long time and devouring the new nature as much as enjoying his newly gained freedom, but a future for humankind. For many species. For the planet. He wanted to scream.

But not all is well that ends well. Soon he would find out the price that had to be paid. Not only the prison sentence. Not only 20 years of his young life. Not only the death of one of his fellow rebels. The biggest price had to be paid now and had to be paid by everyone: the loss of human freedom. It would soon turn out that the electronic device implanted under his skin was not just there to track him and control his behavior in the prison, something he expected they would take out when he left. The device was compulsory for everyone. With the help of electronic technologies, the entire city was transformed into something like a giant prison. A prison without physical walls, but a prison nevertheless. It had been argued that, since most humans were too stupid and lazy to self-regulate, the only solution to stop environmental degradation, climate disaster, and mass extinction was to directly control, manipulate, and coordinate human behavior. People, so it had been argued, need to be forced to do the right thing since they failed to change their lifestyles, threatening their own survival. People need to be saved from themselves. And since human leadership had proven itself to be highly unreliable and toxic (as became all too clear in times of crisis such as the Coronavirus pandemic of 2020) and was utterly unable to deal with complex climate problems given the enormous amount of data and the complexity of the system, it had been decided to have artificial intelligence (AI) take over control of the planet and make decisions instead of humans. Policy decisions, but also many individual decisions, were now delegated to machines. The soft female voice that regulated his use of water, his food choices, and his use of transportation would also accompany him outside the prison. The system would nudge him if possible, but enforce if necessary, and in any case always track and monitor. Very much like the prison from which he was just released. An electronic panopticon was created, building upon earlier achievements by the private sector via social media. But, so it had been decided, this was the only solution for dealing with global risks and vulnerabilities. Humans could no longer be trusted. After the green revolution, AI was given absolute power. In the interest of the planet and in the interest of humanity.

Yet this was difficult to accept for our prisoner. When he realized that the AI that monitored him in the prison and took care of his well-being would never leave him, and when he found out about the new situation in which humanity got itself, he was no longer hopeful or triumphant. Is this what he and his group wanted when they risked their lives in their protests and in their terrorist activities in order to save the planet? Does this justify the bloodshed of the uprising? Is this the ideal his friend died for? She was so young, and such a wonderful person. His mood sunk low when he thought about her. In the train on the way to his new life, he started questioning everything. He tried to remember the old world. He tried to get his head around the problem that people's freedom had been traded for a green utopia, a Brave New World that turned out to be a new prison. Suddenly he felt the weight of the 20 years again, and when he looked out of the window the image of a burdened, pale middle-aged man stared back at him. The smile was gone.

Soon his train of thoughts and worries would be interrupted by a friendly voice that appeared to be concerned with his mental health and well-being, and that would be his guide and guardian for the rest of his life. He would always remain a prisoner.

Introduction: Two Hot Topics Combined, the Question Concerning Freedom, and the Aim of the Book

AI and climate change are both hot political topics, and how they relate and what kind of political challenges that relation raises needs more thought and discussion than it has received so far.

On the one hand, there is much excitement and fear about the new possibilities of AI. The availability of a large amount of data (big data), enhanced processing power, and advances in machine learning and natural language processing have led to smarter software and more autonomous systems, with applications in many areas such as medical diagnosis, translation, and autonomous transportation. The use of machine learning algorithms seems especially promising. As AI gets better and leads to spectacular successes, some dream of general AI that is human-like. Some stress the potential benefits; others fear that AI will take over and that human beings will be the slaves of AI – perhaps merely their raw materials. Suddenly, or so it seems, technology becomes a political question: Who is in control? And what is AI and related technology already doing to us, for example, when we use internet-based services? Are users exploited for their data? Are we increasingly becoming the slaves of our devices and the companies that create the apps we use? Are we slowly but surely moving toward a surveillance society? Is AI threatening democracy? So-called 'digital authoritarianism' seems on the rise, as a report from NGO Freedom House suggests (Shahbaz 2018). It is mass surveillance led by the state: Big data and credit scoring help to nudge the behavior of citizens and other actors. China is already doing this with its infamous Social Credit System, a reputation system that keeps records of individuals to track and evaluate their trustworthiness, which is linked to other surveillance systems that use facial recognition and analysis of big data by AI. Is this also the future of Western democracies? And have we *already* traded some of our freedom for convenience and security, for example, when we mindlessly discharge our data to social media companies and uncritically accept extensive security control at airports?

On the other hand, there is much to do about climate change and what is increasingly known as the climate crisis. Rising greenhouse gas levels contribute to rising global average temperatures and the earth's biodiversity is decreasing rapidly. There are also other huge environmental challenges which seem related to climate change such as air pollution, ocean warming, and extreme weather events. All this has become not only a scientific but also a huge political topic. At least partly due to climate activism by, for example, the school strike movement, and Extinction Rebellion, climate change is now high on the political agenda at all levels, including the global one. Here, too, one of the political challenges concerns freedom: if our predicament is really that bad, should we leave people, corporations, states, and other actors free with regard to what they might (not) do about climate change, or should we enforce action, and, if so, in what way? Should authoritarian states such as China be the model for how to act, given that state-controlled action may be a more effective way of dealing with the climate crisis? Or should libertarian politics such as in the US be the model, largely leaving action up to individual actors themselves, thus preserving a high degree of freedom but without guarantee that there will be change – or only change for the worse? Is there a middle way? How much freedom can we afford in times of crisis, and how much unfreedom can and should a liberal-democratic system tolerate?

How do AI and climate change relate, and what is the political dimension of this relation, especially in terms of freedom? AI may harm the environment and contribute to more global warming, for example, by increasing energy consumption of data centers and by requiring more of the earth's resources for producing smart devices. Consider, for example, the environmental impact of big data initiatives, which raises and ethical and political questions (Lucivero 2020). AI also has a material, physical dimension and this has an environmental footprint and impacts climate change. But AI and data science can also help us deal with climate change, for example, by enabling smart and clean transportation, air monitoring, and early warning systems. A report by PWC (2018) identifies a number of AI applications that may help us to deal with climate change challenges, including smart grids, optimized traffic flows, precision agriculture, and smart recycling. More generally, it is clear that AI is not just a threat but may also help to achieve more sustainability. For example, at the global level it may contribute to the UN's Sustainable Development Goals (Vinuesa et al. 2020). Yet it is not clear if such innovations will be sufficient for dealing with the climate crisis. What if companies' insistence that they will develop AI for 'the earth' and use AI in a sustainable and climate-friendly way is just 'ethics washing', a fig leaf for doing business as usual? What if they do change, but the effect is very small? What if little happens because effective political institutions for global governance of the climate (and AI) are lacking, that is, what if there is too much freedom? And what if AI-enabled measures to mitigate climate change are combined with authoritarian and freedom-threatening tendencies? Implementing the idea that AI will save the planet may well become very problematic when it comes to freedom.

Consider the following scenarios. If nothing is done in terms of regulation, corporations may use AI to analyze people's data, manipulate people toward more consumption, and enhance the efficiency of exploitation of the earth, thereby continue damaging the environment and the climate but also human freedom. People would then be used as mere means for corporate profit, without respect for their freedom. They would be reduced to data cattle and consumption machines. To some extent this is already happening today. As Marcuse and many post-war thinkers already taught us, so-called 'free' democracies are not invulnerable to the rise of totalitarianism and, even if they do not become totalitarian, they may develop new forms of domination and unfreedom. If, on the other hand, the idea is to 'save the planet' and this is seen as a political priority, then radical climate activists or governments that want to implement green policies might argue that AI should nudge us into better environmental behavior, that we should heavily regulate many activities with the help of AI in order to render them more climate-friendly, or perhaps even establish a new kind of authoritarianism with AI in order to ensure the survival of humanity and other species – for example, in the form described in the beginning of this chapter. Again, in this scenario humans would be mere means, but this time toward the end of saving the planet or humanity. In both scenarios, individual freedom is not a priority. Both scenarios use people as means and heavily threaten and damage human freedom. This is ethically and politically undesirable and dangerous, of course, especially if one cares about freedom and liberal democracy.

But what is the alternative, given that there seems to be an urgent need to deal with the climate crisis and other global risks and vulnerabilities, for example, crises caused by deadly viruses and potentially AI itself? How much and what kind of freedom can we have and should have under these circumstances? The leading questions in this book focus on the problem of freedom in the light of climate change and AI: Given the possibilities of AI and given the climate crisis, how much freedom can we afford, what kind of freedom do we want to have, and what are the conditions for freedom? Using resources from the history of political philosophy, such as ideas from Hobbes, Rousseau, Dewey, Marx, and MacIntyre, and connecting to contemporary debates about nudging, the Anthropocene, and posthumanism, this book discusses the question of freedom in the age of AI and climate change. In the course of the chapters, it explores, develops, and argues for a positive and relational conception of freedom and reflects on what this means for our political, technological, and environmental future.

Rationale, Style, Scope, and Context of the Book

This book is not only meant as a contribution to political-philosophical discussions about these topics, but it also aims to help filling what I consider a gap in contemporary philosophy of technology. While philosophers of technology, such as Borgmann, Feenberg, and Winner, have always shown some interest in political philosophy and political theory from the 1980s onward, for example, in communitarianism (Borgman 1989) and in Marxism and critical theory (e.g. Feenberg 1986), in general few have actually engaged with mainstream debates in political philosophy in a sustained way. Even the mentioned authors have prioritized the development of theory within their own field. There has been little exchange between philosophy of technology and political philosophy. Similarly, environmental philosophy has often chosen its own path and – while working in dialogue with other philosophy – has often turned to ethics rather than *political philosophy* when it comes to resources for normative thinking. This book contributes to repairing these shortcomings by mobilizing a key political-philosophical discussion – about liberalism and its stretches and limits – in order to shed light on the challenge of how to deal with climate change and AI. Next to the development of an argument in its own right (about freedom), the book can thus be considered as an exercise in *applied political philosophy* and a kind of 'knowledge transfer' from political philosophy to philosophy of technology and – to some extent – environmental philosophy.

Given this aim, I have chosen a style and scope for this book that is different from that of a targeted intervention in a highly specialized scholarly discussion as is common for, say, an article in a political philosophy journal. Instead, this is a philosophical *essay* on freedom in the light of climate change and AI that is meant to appeal to a wider readership including other disciplines (especially people thinking about technology and environment) and people beyond academia who take an interest in these topics. This does not mean that there is no engagement with scholarly work: although I avoided technical discussions and focused on larger approaches rather than responding to every relevant recent journal article on the topic, it was important for me to show how theory from political philosophy can help thinking about technology and environment. As someone who has a background in both fields, I feel that I am well placed to take this on, although obviously there are limits to my knowledge about political philosophy and limits to the research I could do within the boundaries of this book project. Furthermore, the book has a specific focus when it comes to topics in political philosophy: given that climate change and AI evoke many freedom-related questions in the current public debates, and assuming that most participants in this debate (in the West at least) still adhere to liberalism in some form or other, I have chosen to at least *start* from freedom and liberalism. However, that does not mean that I dismiss other values, principles, and theories. As I will repeat in this book, I believe that other political values and principles are at least *also* important, next to freedom. Perhaps they are even more important. But I limit the scope of this book to a philosophical exercise concerning freedom, which then ends up in discussions showing the *limitations* of liberal thinking and the need for other values, which potentially connects to, for example, Marxist and communitarian thought. For me, this way of proceeding was a more intellectually interesting and challenging project than straightforwardly defending a relational view or arguing for, say, justice or inclusion, since people who reject the latter often appeal to freedom. Moreover, I have chosen to limit my discussion to political philosophy in the Western tradition, since this has enabled me to plug in, and engage with, mainstream political philosophy in Western academia. Since not much work has been done on the nexus of philosophy of technology and political philosophy, and since all textbooks in anglophone political philosophy I found are situated in a Western, liberal-philosophical context, I consider this a good *start*. However, it should not be the end: it would be both interesting and desirable to explore how work in other traditions and other parts of the world bears upon the present topic. Finally, unavoidably there will be omissions or topics that merit further work. I welcome such critical remarks and engagement with the arguments presented here. More generally, I hope this book may contribute to stimulating more research and thinking on the topic – regardless of the particular (sub) discipline and hopefully involving the inclusion of more and different theories and perspectives.

Finally, this book is written during two periods that are often called 'crises': the climate crisis and the Corona(virus) crisis. The first one, which has been called a 'crisis' at least from the time that climate activist Greta Thunberg started criticizing world leaders for not addressing the 'climate crisis' in 2019, but has a longer history in the form of an 'environmental crisis' and is said to be ongoing, has been at least one motivation to think more about climate change and AI, and that alongside my engagement with ethical and political questions concerning AI. How should humanity at large respond to climate change, and what could be the cost of doing so, especially in terms of freedom and democracy? And what is and should be the political significance of unequal distributions in vulnerability to climate change? The second crisis also had an impact on the kind of questions I ask in this book. As someone who has always been worried about the absence of a sufficiently effective political institution for the global governance of global problems, but who *also* cares about liberal and democratic values, the Corona crisis made me even more concerned with humanity's apparent inability to deal with a global crisis and made me at the same time wonder how easy it is for governments to restrict freedoms in a democracy, once a crisis sets in. Finishing this book during a time of lockdown and sensing for the first time in my life at least *some* risk of authoritarian rule and totalitarian surveillance, as when governments set up a curfew or use of data from smartphones to deal with the crisis, it seems to me that our political institutions and values are at least as vulnerable as the humans, bodies, and natural environments on which they ultimately depend.

Overview of the Book

The book is structured around a number of ways in which dealing with climate change and using AI threaten freedom and in the end challenge liberal and liberal-democratic thinking to stretch itself: to broaden what it means by freedom, or to fabricate or appeal to concepts, principles, and values outside of the classical liberal tradition. These discussions mobilize a number of questions, discussions, and metaphors from the history of political philosophy. Consider well-known political-philosophical metaphors such as Leviathan, the Grand Inquisitor, and the Invisible Hand. And of course they will make use of thinkers who have something to say on freedom and the political, such as Hobbes, Rousseau, Berlin, Dewey, Arendt, and MacIntyre. The book thus employs material that is not often used in contemporary discussions about AI and other new technologies. It also responds to more recent debates about nudging, the Anthropocene, and posthumanism – again bringing together unlikely intellectual companions. And adding Haraway's posthumanism to the discussion means that more unusual metaphors also enter the stage: biblical monsters, medieval horror figures, and gothic fantasies are joined by Gaia and, believe it or not, a compost pile. In the course of the book, I will treat metaphors not as a kind of 'illustration' of 'serious' concepts that can and should be carefully disentangled, but as *central* to philosophical work: I will assume that the use, examination, and critical discussion of metaphors belongs to the core business of philosophy, including political philosophy. The result of the book, then, is not only an introduction to the topic of the book and the questions it raises, but also the development of an argument about the meaning of freedom: in the end, this exercise leads to a proposal for a revised, more positive, and relational conception of freedom. This conception of freedom has implications for the role AI and data science can play and should play with regard to dealing with climate change and other global crises, and may also be of broader interest to researchers in political philosophy and related fields.

Note that in the course of the book I will use the terms 'freedom' and 'liberty' interchangeably. As always in philosophy, it matters a lot how the terms are defined. And as I have explained, this is precisely what this book is all about: what does political freedom/liberty mean in the light of AI and climate change?

Let me give an overview of the chapters:

In this chapter (**Chapter 1**) I have introduced the theme of this book by offering a mini story about green authoritarianism powered by AI. Then I have shown how this book connects two hot topics by asking the question concerning freedom: freedom seems threatened in a world of AI and climate change. The book responds to this problem: I have explained the aims, rationale, style, and context of the book. Now I continue with an outline of the other chapters.

Chapter 2 follows up on the green Brave New World idea from the story in the beginning of the book. The challenge is that, in order to deal with climate change and the existential risks of AI, it may well be necessary to have an authority, perhaps even a global authority such as a world government, which has the power to take the necessary measures. To use Hobbes's metaphor: it looks like we need a 'green Leviathan.' If the survival of humanity and human civilization is at stake, it seems justified to create a world order that is able to effectively deal with the problems and indeed 'saves the planet'. But this Hobbesian problem definition and solution seem to lead to authoritarianism: there is at least a very slippery slope toward it. Moreover, the focus on the problem of freedom excludes questions regarding knowledge/expertise and ethics. For example, Plato thought that rulers need to be wise and virtuous. What is and should be the relation between politics and the good life (and the good society)?

Chapter 3 first considers using the notion of nudging, popularized by Thaler and Sunstein: what if we nudge people into better environmental behavior by changing their choice architecture? AI and data science can help to manipulate people in this way by analyzing large amounts of behavioral data and by offering recommendations that nudge people into making more climate-friendly choices. But this recipe presents again problems for freedom. How 'libertarian' is so-called libertarian paternalism really? Deepening the debate, the chapter connects the discussion about nudging to that about human nature. Hobbesian thinkers, utilitarian social reformers, and 'libertarian' proponents of nudging share a pessimistic (and patronizing) view of human nature, not unlike that of the Grand Inquisitor in Dostoevsky's story. People are not able to carry the burden of freedom and are too stupid to make good use of their freedom; it is therefore better that someone else who knows better tells them what to do, for their own good. This contrasts with Rousseau's view of human nature – human nature is essentially good – and his non-Hobbesian conception of freedom as self-rule. However, in the form of the general will, his proposed solution to the risk of authoritarianism creates its own problems. Rousseau also raises the question concerning education: perhaps we should expect more in terms of knowledge and morality not only from rulers, as Plato argued, but also from citizens, especially if we want them to self-rule.

Chapter 4 further explores non-Hobbesian definitions of the problem of freedom and asks what kind of freedom we want. The previous chapters focused mainly on what political philosophers call 'negative freedom': freedom from external constraint. But with Rousseau we also touch on the controversial concept of positive liberty. This can be defined in terms of self-rule (Rousseau) or self-mastery (Berlin), but it can also mean asking the question: once I have negative freedom, what can I do with my freedom? This question leads us to considering what Sen and Nussbaum have called 'capabilities'. Perhaps doing good for the planet and people can be defined in a more relational way: not something that turns me into a means for a higher purpose or that potentially threatens my freedom, but (also) something that increases my positive freedom as a being whose capabilities crucially depend on others and on the environment and the planet's ecosystems. The notion of positive liberty makes us think about the conditions for liberty, and raises questions concerning the relation between, on the one hand, (individual) freedom and, on the other hand, collective good and ethics.

Chapter 5 offers more support to the point that freedom is not the only important political principle and that there are also collective problems and solutions. First it connects problems regarding climate change and AI to that of the so-called 'Anthropocene'. AI seems part of the kind of collective hyper-agency humanity developed vis-à-vis nature and the planet, increasing its geological and climatological impact. It looks like we are ransacking our own home, as Crutzen and Schwägerl put it. How can we conceptualize this kind of collective agency and solve the problem of the Anthropocene? I discuss this by evoking Adam Smith's metaphor of the Invisible Hand. The metaphor is usually used in justifying neoliberal laissez-faire economics, but we can also employ it to discuss collective agency at the planetary level: it seems that there is an invisible hand that leads us to climate disaster. But what happens if – with the aid of AI – we lift the veil and realize that there are many visible hands contributing to our predicament? What if it turns out that some hands do much more than others? Moreover, it seems that we need collective and collaborative action to solve the global problem. Or, questioning solutionism and heeding Heidegger's advice, do we rather need the opposite of doing?

All the same, discussions about collective action still assume that we find ourselves in a Hobbesian situation. In the later part of the chapter, I explore what it would mean to move beyond Hobbesian problem definitions and beyond the idea that freedom is the only political value (thus beyond liberalism understood as radical libertarianism). Politics should not just be about survival and freedom; there are also *other important political values and principles*, for example, justice. For example, what does climate justice mean? One could understand it as requiring a redistribution of climate risk and vulnerability. What are the political consequences? Influenced by Marx we could ask: Will the climate proletariat stand up against the ruling elite and the injustice they create, for example, when smart data analysis clearly shows who is (dis)advantaged and who creates these (dis)advantages? And what does self-rule mean if we find ourselves in the hands of forces we no longer can control?

Chapter 6 questions what seems to be an anthropocentric bias in the theories and discussions construed, discussed, and performed so far. It explores the 'dangerous idea' that freedom and the politics of climate change and AI is not only about humans but also about non-humans. For this purpose, it reviews theories offered by contemporary political theory, in particular Donaldson and Kymlicka's argument for giving citizenship to animals and posthumanist theory by Haraway and Latour. Each in their own way, these authors question the Aristotelian assumption that only humans are and can be political subjects, agents, and patients. What if we admit animals to the political realm, given that they also have needs, interests, and capabilities? What if we enlarge the collective to include non-humans – perhaps given a voice by humans, as Latour proposes? Is this already happening in debates about climate change? What does it mean to build a hybrid 'body politic'? And what if we replace the Leviathan monster by a more diverse, 'tentacular', and relational one, as we could propose using Haraway? (This is the part when Gaia and the compost pile enter the stage.)

These posthumanist directions invite us again to take seriously one of the driving ideas of this book: climate change and AI are not just 'environmental' or 'technical' matters that are merely objects of (human) politics but are deeply political themselves. And if we really want to think about the future of the planet, not only humans and their freedom count. For the question regarding freedom, this means that we should at least also consider the freedom of non-human animals, an exercise which may not only be good for those animals but could also help us to further reflect on our own human freedom. Furthermore, in the Anthropocene we have to rethink what we mean by 'nature'. Given that there is no such thing like an entirely external 'nature' but that we as humans continuously and significantly transform the world, I suggest that instead of a control-obsessed Leviathan we may rather need a politics of engagement and care. I also argue that MacIntyre's relational and communitarian view of ethics and human being can be interpreted as a view of positive freedom: we can only be really free if we can develop human flourishing through dependence on others and in community. This also leads us to consider a developmental and rational approach that sees humans, their freedom, and the environment on which all this depends in terms of *becoming*. Finally, like all our political institutions, democracy is vulnerable and mortal; this is an ongoing challenge but also gives us a chance to transform it – preferably in a way that takes into account that politics in the 21st century is not only about humans but also about (non-human) animals, climate, viruses, and AI. Boundaries are shifting and probably should be shifted. The body politic, as a technological artifact, 'is' not; rather, it *becomes* and needs continuous renegotiation and rebuilding. This makes politics and liberation into a 'poetic' project.

The book ends with a chapter that summarizes the challenges for freedom, liberal democracy, and liberal-philosophical thinking identified in this book, and offers some (normative) conclusions. **Chapter 7** argues that in times of crisis, when we try to deal with the new and global vulnerabilities and risks related to climate change, AI, pandemics, and so on, we have the choice between either destroying liberty altogether – which, unfortunately, sometimes even happens in the name of freedom –

or finding new balances with other values, stretching what we mean by freedom, and revising the liberal-democratic model and idea of politics in ways that enable us to deal with the new problems. The challenge with regard to freedom and liberalism for those who dare to call themselves liberals is: stretch it before others break it. I conclude from the previous chapters that this stretching leads us into a more relational direction: it means accepting that realizing and maintaining freedom and liberal democracy depends on creating the conditions for their flourishing, and that this requires supporting capabilities, realizing human flourishing through dependency and community, and helping to realize the common good and the good society, next to – in times of crisis – securing the survival of the collective. Stretching also means acknowledging that there are other values that are important next to liberty, such as justice and equality, and opening the collective to include non-humans.

I will conclude that, in principle and in the light of the problems raised by AI and climate change, the conception of negative liberty is inadequate and even dangerous if not joined with a positive one. Next to ensuring the survival of the collective at the global level through collective action and international cooperation, we need to act on a positive and *relational* understanding of freedom which promotes capabilities, acknowledges dependency, and links freedom to its social and environmental conditions; a merely negative and individualist conception does not enable us to flourish and ultimately will threaten our survival as humans and humanity. AI can help us to create these better conditions, but only if we understand that technology and its future is bound up with the future of humans and their freedom – which in turn is all about the future of non-humans, the environment, and 'the planet'. At the end of the book, I call for the use and development of AI and other technologies that support the growth of this relational kind of freedom, which is part of our responsibility as one of the main makers and poets of the Body Politic. I call for *sympoietic* technologies and sciences that, integrated in this positive and relational poetic-political project of liberation and democratization, contribute to the making of a more inclusive collective and the building of a new common world.

References

Borgmann, Albert. 1989. "Technology and the Crisis of Liberalism: Reflections on Michael J. Sandel's Work." In *Technological Transformation*, edited by Edmund F. Byrne and Joseph C. Pitt. Philosophy, 105–122. Dordrecht: Springer.

Feenberg, Andrew. 1986. *Lukács, Marx and the Source of Critical Theory*. Oxford: Oxford University Press.

Lucivero, Federica. 2020. "Big Data, Big Waste? A Reflection on the Environmental Sustainability of Big Data Initiatives." *Science and Engineering Ethics* 26 (2): 1009–1030.

PwC (PricewaterhouseCoopers). 2018. *Fourth Industrial Revolution for the Earth: Harnessing Artificial Intelligence for the Earth.* PwC. https://www.pwc.com/gx/en/sustainability/assets/ai-for-the-earth-jan-2018.pdf

Shahbaz, Adrian. 2018. "The Rise of Digital Authoritarianism: Fake News, Data Collection, and the Challenge to Democracy." Freedom House. Accessed 30 March 2020. https://freedomhouse.org/report/freedom-net/2018/rise-digital-authoritarianism

Vinuesa, Ricardo, Hossein Azizpour, Iolanda Leite, Madeline Balaam, Virginia Dignum, Sami Domisch, Anna Felländer, Simone Daniela Langhans, Max Tegmark, and Francesco Fuso Nerini. 2020. "The Role of Artificial Intelligence in Achieving the Sustainable Development Goals." *Nature Communications* 11 (233). doi:10.1038/s41467-019-14108-y

Figure 2 Frontispiece of book *Leviathan* by Thomas Hobbes: Engraving by Abraham Bosse.

2 Leviathan Reloaded

Climate Change and AI as a Hobbesian Problem for a Vulnerable Civilization

Introduction

The story at the beginning of this book is based on the idea that two interlocking developments – the climate crisis and progress in the area of AI – raise the problem of the political, in particular the problem of the political as formulated in the Western modern philosophical tradition: the problem of *freedom* and the related issue of the tension between individual freedom and the good of the collective. This is a long-standing problem, but now it re-emerges in a new context: this book discusses it in the light of the potential uses of AI and the possibilities for dealing with the climate crisis, in the hope that such a discussion can then be used to create and critically discuss scenarios and visions for our global future.

So, let us begin: what, exactly, is the problem concerning freedom? Let us look at some problem formulations and their corresponding solutions, drawn from the history of political philosophy. I start with an idea that was also developed during a time of crisis, namely, the English Civil War: Leviathan. Then I will introduce the (much later) concept 'Tragedy of the Commons' and ask further questions that link the problem of freedom to questions concerning knowledge, ethics, and democracy. This will constitute a first and brief attempt (essay) to explore the limits of freedom and challenges for liberal thinking in the light of AI and climate change – an exercise that will be continued in the next chapters.

The State of Nature and the Call for a Monster or Mortal God to Solve the Problem

Today most people in the West would say that they prefer a liberal-democratic way of dealing with new technologies, meaning that these technologies should be developed in a way that does not threaten freedom and democracy. And in spite of many science-fiction films that suggest the opposite, even some people who think about the future of technology are positive that we can do this. For example, in *Citizen Cyborg* (2004), James Hughes has argued for technologies that enhance humans but in a way that preserves freedom and democracy. He embraces the Enlightenment ideals of liberty, equality, and solidarity and thinks they could power a version of transhumanism (the project to re-design the human). This suggests that technologies such as AI can be successfully embedded in a liberal-democratic institutional framework, at least provided that we do care not only about freedom but also about equality and solidarity (I will return to this point later).

Others are less optimistic. For example, Paul Nemitz (2018) has voiced the concern that AI will threaten democracy because of digital power accumulation, and Harari (2015) thinks that democracies cannot cope with AI and biotechnology, which 'might soon overhaul our societies and economies' (375). He believes that 'the liberal story is flawed (...) and that in order to survive and flourish in the 21st century we need to go beyond it'. We are not free individuals but 'hackable animals' (Harari 2018). But even if we look at the present situation, it becomes increasingly clear that relatively new technologies such as the internet and AI do not necessarily lead to freedom and democracy; they also turn out to be ideal instruments for authoritarian regimes and rulers in 'democracies' with authoritarian tendencies. But what is the alternative? Is absolute freedom with regard to these technologies better? Is it better to have no governance at all for AI and advanced technologies? What we see today is that, especially at a global level, there is very little leadership and co-operation, let alone robust governance, with regard to technologies such as AI. Thus, there seems to be a choice (dilemma) between two evils: authoritarianism and radical libertarianism.

Similarly, dealing with climate change is not necessarily a story of liberty and democracy, and currently there is very little governance in place at a global level. Here too, most people would again declare their adherence to liberty and democracy. In his book *Green Liberalism* (1998), Marcel Wissenburg already argued that liberalism as a political philosophy is capable of absorbing green ideas and ideas around sustainability, even if this means that some limits to freedom have to be accepted (e.g. limits to property rights and limits to abuse of natural resources). But it is unclear how liberalism fares when it is under pressure from the climate crisis. It seems that to tackle this problem, much more radical measures have to be taken, measures that go more in the direction of authoritarianism since they would imply more severe limits to freedom, for example, regulating the behavior of individuals and organizations. How can liberalism survive *this* crisis? Will the climate crisis give liberalism its final blow? And what if, on the other hand, at the global level there is total *lack* of governance and coordination, as turns out to be the case today? How can we deal with the climate crisis if there is more or less a laissez-faire situation? We face a dilemma similar to the one raised by AI: the choice seems to be between absolute freedom – which preserves liberty but at the cost of not dealing with the problem at hand – and absolute authoritarianism, which solves the problem but at the cost of liberty.

This dilemma is not unknown to political philosophers. One could argue that the global situation concerning the potential uses of AI and the governance of climate change looks very much like what the 17th-century philosopher Thomas Hobbes calls a 'state of nature'. In this book *Leviathan* (originally published 1651), he describes a state in which there is absolute freedom and no government. Applied to the climate crisis and AI, a comparable state of nature means that individuals decide for themselves what to do about climate change and AI; while there might be governance in other areas, *when it comes to AI and climate*, there is only private judgment, policies, and action. This could also be applied to corporations and states and at the global level: each makes its own decisions about AI and climate change, without a central authority regulating them. They try to preserve themselves; they try to survive and compete.

And this leads to troubles. According to Hobbes, a state of nature without political authority is a state of competition and conflict, in particular conflict about resources. Since humans seek gain and reputation, there is violence and war: a war 'of every man, against every man' (84). People experience 'continual fear, and the danger of violent death' and humans' lives are 'solitary, poor, nasty, brutish, and short' (Hobbes 1996, 84). Some of this we see today when states (nearly) fight about the remaining oil, when each country makes their own regulation for AI, and when each corporation has its own ethical AI principles. At a global level, there seems to be a kind of state of nature: a brutal race for AI without global governance and, in the light of climate change, conflicts between nations and between people for survival in a condition that is likely to get increasingly nasty.

With regard to the climate crisis, one could add that there is a 'Tragedy of the Commons', an idea from 19th-century economist W.F. Lloyd, who argued against Adam Smith: individuals acting according to their self-interest leads not to the common good, as Smith argued, but instead to depletion of shared resources. Lloyd's example was grazing on common land (the common): if unregulated, each herder would let so many animals graze in order to maximize their advantage, which then leads to a depletion of the common. The solution is regulation. Similarly, in discussions about the environment and climate change, the concept is used to defend sustainable development and doing something against global warming: without regulation, the earth and its ecosystems and resources are used as commons and will be depleted and destroyed. In 1968, the American ecologist and philosopher Garrett Hardin already argued that when resources such as oceans, rivers, and forests are used without anyone taking responsibility and managing, this leads to depletion. Considering such resources as free is part of the very problem. Later in 1987, the Brundtland Commission noted in its famous report that there is a similar problem at the global level, calling for management of the commons on the basis of a rather Hobbesian assessment:

> The Earth is one but the world is not. We all depend on one biosphere for sustaining our lives. Yet each community, each country, strives for survival and prosperity with little regard for its impact on others. Some consume the Earth's resources at a rate that would leave little for future generations.
>
> (World Commission on Environment and Development 1987, 39)

The report complains that traditional forms of national sovereignty are particularly problematic when it comes to managing the 'global commons' and their shared ecosystems; this requires 'international cooperation and agreed regimes for surveillance, development, and management in the common interest' since all nations depend on the commons (WCED 1987, 258). Extending the report's definition of commons from areas that fall outside national jurisdictions to the planet as a whole and its atmosphere, the pollution and over-exploitation of which threatens us all and creates a Tragedy of the Commons, one could call for (more) international cooperation and global management of the planet.

Thus, the problem is that at a global level we face a Hobbesian and 'Tragedy of the Commons' type of situation, with nationalism being a barrier to change. While the 'Tragedy of the Commons' concept has its limitations with regard to dealing with climate change – instead of regulating we may want to stop 'grazing' altogether when it comes to extracting and using fossil fuels, for example – it is easy to argue that we are (still) in a kind of Hobbesian situation where no governance of the earth and shared ecosystems leads to tragedy – including climate tragedy.

How do we remedy this situation? Hobbes imagined that individuals would make an agreement, a social contract, to submit to an absolute sovereign authority and power that brings peace and protection. Individuals accept a restraint upon themselves in order to get out of their miserable condition and preserve themselves. He called this authority a Leviathan, a monster, but one that is necessary to preserve the peace. It is a 'Mortal God' (Hobbes 1996, 114) which can produce 'the terror of some power' by means of word and sword, thus delivering security (111). This then enables survival of individuals.

Given our interest in AI, it should be noted that Hobbes's metaphorical toolkit includes not only the biblical monster in the form of a sea serpent (Hobbes borrowed Leviathan from the book of Job), but also automata. When he describes Leviathan in the very beginning of the book, he uses the automaton as a metaphor. Leviathan is an 'artificial man', imitating the human. Hobbes compares it to automata that move by themselves and thus have 'an artificial life'. Leviathan or 'Commonwealth' or 'State' is

> an artificial man; though of greater stature and strength than the natural, for whose protection and defence it was intended; and in which, the sovereignty is an artificial soul, as giving life and motion to the whole body.
>
> (7)

Leviathan is thus an artificial 'body politic' – in contemporary language: a cyborg body politic or robot body politic – consisting of magistrates, executioners, and so on, created by the pacts and covenants that make up the social contract. Leviathan is a cyborg body created by social agreement. Hobbes compares the nature and performance of such agreements to the divine *fiat*, the word pronounced by the Jewish-Christian God that performatively creates the human and everything else that is natural. Leviathan is an artificial construct, an automaton, that is created, not by a god, but, by humans using words. To use a 20th-century term from philosopher of language John Searle: Leviathan is created by linguistic declaration (Searle 1995, 2006). This *artifact* then ends the brutal, nasty competitive and violent condition.

Inspired by Hobbesian reasoning, one may call for a green Leviathan or climate Leviathan to take up the role of the sovereign at the global level in order to ensure regulation that prevents the perishing of collectives and the deterioration of the natural environment that is a condition for their survival (and by implication also their flourishing – but for now I limit myself to the Hobbesian framework). We would collectively – as humanity – decide and agree to have such a monster regulate us. A cyborg body politic at a global level could help us to deal with climate change in an effective way and ensure the survival of humankind. And we could do the same to solve the global 'state of nature' situation when it comes to the governance of AI. For example, such a Leviathan would prevent the use of AI for war purposes and other unregulated behavior and restrain boundless competition. Since self-restraint on the part of individuals, corporations, and nation states does not work, so this argument goes, it is better to have an authority regulate – at all levels. The result of the agreement is a Leviathan since it is an artificial political construct, a Hobbesian automaton, which we decide to bring into being by means of agreement (words) and which has sufficient authority and strength to make sure that the climate and environmental crisis is dealt with and that AI is properly governed in order to secure survival and peace as individuals, states, and humanity. If we don't do this, the Hobbesian argues, things will get nasty.

Note that this global Hobbesian situation and the related authority and coordination problem may also be created by (risk of) natural disasters such as a pandemic and of course by other technologies. In his recent article, 'The Vulnerable World Hypothesis' (2019), Nick Bostrom, another transhumanist philosopher, has warned that science and technology may destabilize civilization. Global governance is then a way to stabilize a vulnerable world, 'one in which there is some level of technological development at which civilization almost certainly gets devastated by default' (Bostrom 2019, 455). There could always be a technology – Bostrom uses the somewhat odd and potentially problematic metaphor 'black ball' – that destroys civilization. As he also suggested in previous work when he wrote about existential risks, AI could be such a destructive force (467), putting humanity at large at risk. But nuclear weapons or biological weapons could do the same. In the absence of global coordination, civilization will end. In what Bostrom calls the 'semi-anarchic default condition (…) There is no reliable mechanism for solving global coordination problems and protecting global commons' (457). Like Hobbes's Leviathan, global governance is then proposed as a kind of necessary evil to ensure the survival of our civilization, since collective action is needed to deal with the problem:

> The set of state actors then confronts a collective action problem. Failure to solve this problem means that civilization gets devastated in a nuclear Armageddon or another comparable disaster. To deal with (this type of) vulnerabilities, what civilization requires is a robust ability to achieve global coordination.
>
> (Bostrom 2019, 467)

Bostrom's argument also seems to apply to global warming: once a tipping point has been reached (and many people today argue that this is already the case), collective action in the form of global governance seems the only way we can deal with this problem. The alternative is global disaster and perhaps the end of civilization as we know it. We need the monster, we need a new robotic body politic. (Note that this assessment differs from Bostrom's own view, which suggests that presently global warming is not dangerous enough and instead asks us to imagine situations 'with far greater civilization-destroying potential than the actual condition', for example, a temperature rise of 15 to 20 degrees Celsius (460).)

As Bostrom would acknowledge, the Hobbesian solution is problematic in terms of freedom, since people are ruled by an absolute authority. But there is a further problem. The Hobbesian automaton or mortal God does not only have the divine-like quality of omnipotence (all powerful); to keep the order, the monster also has to see everything. Hobbes quotes Job 41: 'Hee seeth every high thing below him'. This could be interpreted as a call for surveillance, necessary for the exercise of quasi-divine power. Later, in the 18th century, Jeremy Bentham proposed the ideal of a panopticon: a prison in which all prisoners could be watched at all times but the prisoners would not know if they are watched. Today, AI and other information technologies risk putting an entire society under surveillance: a 'surveillance society' (Lyon 2001), a 'control society' (Deleuze 1992), or what Shoshana Zuboff (2019) has called 'surveillance capitalism': an economic order that uses human experience as raw material for commercial practices of extraction, prediction, and sales, and in which machines are not only used to dominate nature but also human nature: behavior of individuals, groups, and populations are modified in the service of market objectives (Zuboff 2019, 515). While not all surveillance necessarily takes place top down (there is also peer-to-peer surveillance and self-surveillance) and while the question of power is more complicated than this (see, for example, the work of Michel Foucault), clearly today not only authoritarian rulers and their Hobbesian automata but also powerful corporate players have plenty of instruments available for surveillance and manipulation.

AI is such an instrument, of course. It is part of the problem and part of the solution: it raises the Hobbesian problem and at the same time it can be used to solve such problems by enabling control, manipulation, and surveillance. For example, Bostrom proposes not only world governance but also predictive policing powered by ubiquitous surveillance in order to stabilize vulnerabilities (Bostrom 2019, 467). While he acknowledges that this could help despotic regimes and 'enable a hegemonic ideology or an intolerant majority view to impose itself on all aspects of life' (468), he also enumerates some good consequences of surveillance and world governance: next to the stabilization of civilizational vulnerabilities, it could lead to reduction of crime, prevention of war, solutions to environmental problems, and social transfers to the global poor (468). But while a solution such as global governance and surveillance has its own risks and benefits, the main – Hobbesian – point is that it can effectively deal with the problem. Via collective action and global coordination, it is able to reduce civilizational vulnerability with regard to AI, climate change, and so on, and thereby prevent global disaster and ensure the continuation of human civilization. Bostrom's scenario can thus be helpfully interpreted as constituting a Hobbesian argument, which justifies installing a central authority by saying that it is needed to coordinate and, in this way, ensure survival, security, and peace.

Beyond Hobbes

Between the Horns of the Dilemma?

But do we have to choose between absolute liberty and absolute authority? One could try to go between the horns of the dilemma and accept *some* constraints to liberty, without going all the way to authoritarianism.

This is in fact what some societies try to do, for example, European states influenced by social democracy. However, such constraints to liberty are difficult to justify *within* liberalism in its libertarian version, that is, by appealing to freedom. It seems that in order to justify this direction, one has to appeal to other values. Or is there another option? Could one *interpret freedom in a different, more creative way*? In the next chapters, I will take up both (non-Hobbesian) suggestions.

Another alternative is of course to explore other political-philosophical traditions. For example, when we look at the republican tradition in political philosophy (not to be confused with the US 'republican' party), there are some interesting questions and elements that we can add to the liberal-contractarian discussion, questions and elements that are very relevant in the light of the problems under consideration (AI and climate change).

Knowledge, Virtue, and Wisdom

First, in the Hobbesian view there is no requirement that the authority has *knowledge and expertise*. If the authoritarian ruler does the job – that is, keeps the peace, keeps the social order, then there is nothing more expected. The sword is enough. This is different in the republican philosophical tradition starting with Plato. In the *Republic*, he also proposes authoritarian rule, but according to him, knowledge is required. He argues that it would be 'absurd to choose anyone but the ones who have knowledge' (484d), in particular philosophers who love the truth (485c) and love wisdom (485cd). He proposes an aristocracy ruled by a philosopher-king (473d), who doesn't love money (485e) but has knowledge. In Plato's view, this is a particular kind of knowledge: knowledge of the forms of justice and goodness. He or she (Plato/Socrates thinks women can rule too) is wise or at least has the love of wisdom, is virtuous, and is selfless. Plato's philosopher-king (or queen) leads a simple life. And he or she is crafty, has know-how. He compares the ruler to the pilot of a ship: the one who pilots the ship of state must 'pay attention to the seasons of the year, the sky, the stars, the winds, and all that pertains to his craft' (488d). Thus, for Plato the main problem is not a lack of authority but corruption, love of money, and ignorance, which have led to society going into the wrong direction – something which, unfortunately, sounds rather familiar today. Like in Hobbes, the proposed solution is authoritarianism, but this time the sword and the words are guided by wisdom and virtue. The philosopher rules. In this view, there is once again no place for democracy, especially if that means rule of the majority. According to Plato, giving the power to the majority would be giving power to 'madness' and 'savagery' (496cd).

In light of the problem of climate change and AI, the analogy with regard to climate change and AI would be: give power to experts who not only have scientific knowledge but are also wise, virtuous, and good, and therefore know how, and have the legitimation to, steer the planetary ship in the right direction. In this political-philosophical tradition, the point is not the preservation of social order as such and peace keeping (Hobbes), but rather navigation of society into the right direction by non-corrupt, truth-loving, wise, and good rulers who rule in the interest of society as a whole. (Note that of course there is much more to say about Plato's view. For example, for Plato, knowledge consists of more than knowledge of good: knowledge of the forms includes knowledge of what is rather than knowledge of becoming, having had intercourse with 'what really is' (490b), knowing beauty itself rather than many beautiful things, etc. But this goes beyond the scope of the argument I construct here, which is focused on the suggestion to bring in ethics.)

Hobbes does not require any of this from the ruler. Similarly, the story in the first chapter was not about wisdom and virtue. The AI that is used by authoritarian rulers or rules as a Leviathan (one could also say constitutes the head of the Leviathan) has knowledge in the sense of 'artificial intelligence', that is, it can calculate probabilities. It can do statistical analysis on big data. But it has no experience and therefore no wisdom, since wisdom requires experience. It also has no virtue, since it is not a moral agent with mental properties such as consciousness. It is used by humans to manipulate people, but does not itself possess any of the character traits Plato recommends. If humans remain in control of the AI, then, according to the Platonic view, those humans should be wise, truth-loving, non-corrupt people, educated to be philosophers from early age. Intelligence as such, even human intelligence, is not enough. Virtue and wisdom are needed. This contrasts with both the liberty-centered Hobbesian tradition and the contemporary liberal-democratic thinking that grew out of it, which have no such requirement: our democracies do not require that politicians are virtuous or wise. It also contrasts with Hobbesian transhumanism à la Bostrom, which narrowly focuses on intelligence. According to Plato, giving the steering of the ship of state to the savage, wisdom-lacking majority or to a small group of corrupt, power-mongering members of the elite, who only care about money, would be pure madness. And clearly AI or experts who are smart but possess no virtue or wisdom would also not make the cut.

Whereas classical liberalism is 'thin' when it comes to matters of *virtue, the good life and the good society*, the republican tradition is not. In Plato's republicanism there is not only room for a clear vision of the virtuous person, the good life, and the good society; such a vision is *central* and the philosopher-king must have one in order to rule. If his ideas can be interpreted in terms of a social contract at all, it is an agreement to give the power to those who are the most wise, virtuous, and knowledgeable; those who can see what is real, true, good, and beautiful.

Perhaps in the future we will decide that humans are not knowledgeable enough and that therefore an AI (a superintelligence that is more intelligent than humans) needs to take over. This might be a tempting proposal when today we observe so much human stupidity with regard to climate change or when we are impressed by what AI can do. But again, from a Platonic point of view, the problem is not so much a lack of intelligence. Wisdom and virtue are required. And however 'intelligent' AI may be, AI cannot provide this wisdom and virtue. Of course, AI could *help* the guardians and the king to do their job: ensuring the good of society. AI could provide information about correlations and patterns in data. Perhaps it could also be used for steering the behavior of people who lack virtue and knowledge. But it would be up to wise humans (philosopher-kings) to decide and rule, not machines. The human captain (male/female) is in control of the machinery.

Note, however, that Plato himself uses different metaphors. The rulers or the body politic cannot be compared to an automaton (as Hobbes did), since the virtue and knowledge required here cannot be something that can be automated. The political institution is not a machine (work of human beings, which then does its own thing) but is work *by* human beings. The steering of the ship is a craft. And according to the Platonic view a craft is, again, nothing for machines.

Democracy and Authoritarianism

Neither of these proposals is democratic. Both Hobbes and Plato propose an authoritarian solution to the problem of the political. One could and should criticize these proposals and instead endorse a democratic solution: government by 'the people' or 'self-rule'. However, the problem is that when we say this, we do not really know what it means. If it simply means rule by the majority, then this seems very problematic. What if the majority chooses someone who is ignorant or malicious (as is happening today in some countries)? It seems that at the very least other conditions need to be in place for democracy to work. One of these conditions seems to be that at least *some* kind of knowledge and expertise is required. If people rule themselves, then they also need to be knowledgeable and they need to be educated. (I will return to this point.)

When we consider climate change and AI as *global* problems that require expert knowledge to even understand what is going on (let alone solve things), this raises at least the following fundamental challenges with regard to expertise:

First, at the global level the question concerning expertise and governing is not even 'allowed' to arise since there is not even a proper authority (authoritarian or democratic) with sufficient power to tell nation states or multinational corporations what to do. And since in political history there usually has been first an authoritarian ruler and then democracy, rather than democracy growing bottom-up, one could ask if democracy can be established at the global level without first having some kind of authority at that level at all. On the basis of the Hobbesian argument, at least, one would need not just an international organization but a supranational authority with sufficient power to deal with a Hobbesian situation. Democracy seems to presuppose that a Hobbesian solution is in place, but this condition is not available at the global level, and especially not with regard to the problems at hand (climate crisis, unregulated AI). Therefore, there is at least a temptation to put up a global (green) authority, and *then* add other requirements such as knowledge and ethics. As a Hobbesian one could argue that we have to put survival first, even if this destroys freedom.

Second, there is a huge tension between (representative) democracy and expertise, to say the least. Democracy as we know it does not exclude ignorant leaders but at times almost seems to produce them. This is possible since in the liberal-democratic conceptions of democracy there is absolutely no requirement that democratic leaders have any knowledge. The most ignorant can rule, provided they collect sufficient votes. In the light of the surveillance and manipulation possibilities offered by AI, which can easily fall into the hands of idiots and malicious people, and the situation of climate change, which can only be solved by means of knowledgeable and wise action, this is dangerous. We see today that societies such as the US and the UK that claim to promote libertarianism are actually very vulnerable to authoritarianism, including a form of authoritarianism that shuns expertise, academia, scientific knowledge, and truth. Next to having other bad consequences, this is a huge problem for liberty and democracy, for the wise use of technology, and for dealing well with the climate crisis.

Now one could object that it is not democracy itself that produces ignorance but deviation from it. What is missing at both the national and the global level is a lack of democracy, not democracy gone awry. But if we say this, what do we mean by 'democracy'? Again, if democracy is understood as being *merely* the rule by the majority (or those who represent the majority), then democracy itself does not guarantee knowledge and wisdom on the part of those who rule. The same could be said for virtue and corruption. (Nor does democracy as majority rule protect against the rise of authoritarianism, as was already clear at the time when Hitler came to power in Nazi Germany.) So, if we argue that we currently lack proper democracies, then we should add further criteria to the justification of power and the idea of democracy. Inspired by Plato, we could add criteria regarding knowledge and perhaps also character, for example.

What does this mean for the governance of climate change and AI? We could require that any proposed political system that is meant to tackle the problems concerning climate change and AI should meet at least two conditions: it should manage to end the state of nature at the global level (it should solve the Hobbesian problem) and it should not further proliferate ignorance and vice, that is, provide solutions guided by knowledge, wisdom, and virtue (it should solve what we could call, in the context of this discussion, the Platonic problem). But again we must ask: given that both Hobbes and Plato went the authoritarian way, can all this be done in a democratic, non-authoritarian way?

One could argue again – but this time including the issue of knowledge and ethics – that the way the problem has been formulated so far misleads us into thinking that we have to choose between two extremes: a totally authoritarian and totalitarian system (which deals with the problems concerning climate change and AI but is unacceptable from a liberal and democratic point of view), on the one hand, and a totally free democratic system (which cannot deal with the problem and seems to deteriorate into total lack of social order and ignorant leadership), on the other hand. One could argue that this is a false dilemma and that there is a third option (or many third options, in the plural). With regard to knowledge and ethics, such a third option would involve non-totalitarian and non-authoritarian forms of control and manipulation that still preserve some freedom and democratic decision-making, but are based on expertise, wisdom, and virtue. Is this possible? I will need to say a lot more about this, but for now consider the following indications that it is not impossible:

First, nudging as a compromise between libertarianism and paternalist steering is an attempt to retain some kind of freedom and at the same time have expertise flow into shaping of people's choice architectures. The idea of nudging is to change the environment in order to influence people's choices (subconsciously), while leaving them free to choose. It is a non-Platonic option, since there is no virtue on the side of the nudgees; they are manipulated because they lack self-control and (the use of) their rational capacities. Only the nudgers could be construed in Platonic terms, that is, as wise and virtuous philosopher-kings, who know what is best for the nudgees. It is also questionable if it is still libertarian, since here some people (the nudgers) argue that they know better than others what is good for them, and manipulate them for that purpose. One could argue that this is patronizing, paternalistic, and anti-freedom. But it does seem to help with solving the climate crisis. I will say more about nudging and its challenges for liberal thinking in the next chapter.

Second, European democracies that are less free than in the US but more free than, say, China, seem to have taken the 'third' way: they manage to regulate behavior and thus take away some liberty but without having a full-blown totalitarian and authoritarian system, and their policies are based on expertise to at least some extent. This is so at the national level but also at the supranational level: for example, the European Commission consults experts. While there are no explicit expectations regarding the virtue of rulers, expertise clearly plays a role and there is a balance between liberty and other political values. My point here is not that these societies are perfect and 'solved' the issue concerning freedom once and for all (e.g. there may be the danger of bureaucratic authoritarianism), but rather that they show that a third way between absolute freedom and absolute authoritarianism is *possible* and that expertise plays *a* role in their political systems.

However, both options seem to have to appeal to principles beyond the liberal tradition. Freedom, at least in its radical, libertarian form, seems compromised here once, and in so far as, people are ruled by experts or others who claim to know what is best for them. And even if we could agree that freedom is preserved, both options might have a slippery slope toward unfreedom. Libertarian paternalism might become plain paternalism, without liberty, when it moves from subtle manipulation to coercion (or when the former is interpreted as destructive of freedom – see later). Furthermore, the delicate balance between freedom and other values that is carefully maintained in some societies under normal circumstances might be lost very quickly once a crisis hits such a society. Consider, for example, how the Coronavirus pandemic has led rather soon and surprisingly easy to severe restrictions of liberty in European societies. Why would this not happen if climate change were to be taken seriously as a global crisis? As the story in the beginning of this book and the Hobbesian view suggest, it remains a huge challenge to solve global problems related to AI and climate change without significantly eroding freedom. The introduction of the question of knowledge and expertise did not make things any easier. What does democracy mean, once we give experts a larger role and let them decide what is best for us? In sum, what does freedom mean and how much and what kind of freedom can and should we have, once we agree to a new form of social control (made possible by AI), give more power to experts, and agree to give power to an authority in order to deal with climate change and harness the power of AI?

Of course, one could reject the premises and deny the urgency and reality of the problem regarding climate change and AI by arguing that (a) the climate crisis is not so bad as it seems and that (b) AI will not and should not be used for the manipulation of people by governments. One could also argue that there is an urgency concerning climate change and a problem with AI, but that it is up to individuals to change their lives and that we should not change the system as a whole and at a collective level. If this road is taken, then the situation is likely to remain as it is today in many countries and at the global level: one then has to face the consequences of climate change as they come, passively watch ineffective and insufficient action by governments (and no or little international cooperation), leave the manipulative use of AI to private corporations, and hope that individuals will change their behavior. One has to then explain why the current situation and this way of dealing with the crisis will not lead to environmental and human disaster (or why the latter is not bad), why restraining the liberty of individuals, corporations, and states is not at all necessary to deal with the problems we have, and why AI is in good hands with big tech and the rulers we have.

If, on the other hand, one *accepts* the premises, then asking the question of freedom and the shaping of the political (the organization of the collective) is crucial and we need to continue the discussion. If we are serious about tackling climate change and if we want to regulate AI but also use the new opportunities it provides, we have to think harder about what this entails for human freedom and democracy. The discussion in response to Hobbesian and Platonic thinking was a good start, but we have to do more work. In the next chapter, I show that any answer to these questions regarding freedom and democracy will at least partly depend on what the Western philosophical tradition calls the question of *human nature*. I will introduce this problem and its relation to the question concerning freedom by using Dostoevsky's story of the Grand Inquisitor. Then I will further discuss nudging and compare Hobbes's approach to freedom to that of another key thinker in the history of political philosophy: Jean-Jacques Rousseau. This will lead us to a different, equally challenging set of ideas concerning liberty and democracy and thus involve more wrestling with the borders of the liberal-philosophical tradition in the light of the challenges posed by climate change and AI.

References

Bostrom, Nick. 2019. "The Vulnerable World Hypothesis." *Global Policy* 10 (4): 455–476.

Deleuze, Gilles. 1992. "Postscript on the Societies of Control." *October* 59: 3–7.

Harari, Yuval Noah. 2015. *Homo Deus: A Brief History of Tomorrow.* London: Harvill Secker.

Harari, Yuval Noah. 2018. "The Myth of Freedom." *The Guardian*, 14 September 2018. Accessed 17 March 2020. https://www.theguardian.com/books/2018/sep/14/yuval-noah-harari-the-new-threat-to-liberal-democracy

Hobbes, Thomas. (1651) 1996. *Leviathan.* Oxford: Oxford University Press.

Hughes, James. 2004. *Citizen Cyborg: Why Democratic Societies Must Respond to the Redesigned Human of the Future.* Boulder, CO: Westview Press.

Lyon, David. 2001. *Surveillance Society: Monitoring Everyday Life.* Buckingham: Open University Press.

Nemitz, Paul. 2018. "Constitutional Democracy and Technology in the Age of Artificial Intelligence." *Philosophical Transactions of the Royal Society: A Mathematical Physical and Engineering Sciences* 376 (2133): 2018.0089. doi:10.1098/rsta.2018.0089

Plato. 1997. *Republic.* In *Complete Works*, edited by John M. Cooper, pp. 971–1223. Indianapolis, IN: Hackett Publishing Company.

Searle, John R. 1995. *The Construction of Social Reality.* New York: Free Press.
Searle, John R. 2006. "Social Ontology: Some Basic Principles." *Anthropological Theory* 6(1): 12–29. doi: 10.1177/1463499606061731
Wissenburg, Marcel. 1998. *Green Liberalism: The Free and the Green Society.* London: UCL Press.
World Commission on Environment and Development. 1987. *Our Common Future.* (the Brundtland Report) United Nations General Assembly document A/42/427.
Zuboff, Shoshana. 2019. *The Age of Surveillance Capitalism: The Fight for a Human Future at the New Frontier of Power.* London: Profile Books.

Figure 3 Francis Bacon, Study after Velázquez's Portrait of Pope Innocent X.

3 Big Data for the Grand Inquisitor

The Discussion about Human Nature and Freedom from Rousseau to Nudging

The Story of the Grand Inquisitor

In his novel *The Brothers Karamazov* (originally published 1879–1880), Fyodor Dostoevsky tells the famous story of 'The Grand Inquisitor'. In the story, Christ comes back to earth at the time of the Inquisition, which sentences him to death. The Grand Inquisitor argues that humans cannot handle the freedom that Jesus has given them. Freedom of choice is a burden (1992, 255); people live in unrest, confusion, and misery. They 'will never be strong enough to manage their freedom' (261). Human nature is 'feeble' (257). Therefore, he and his colleagues of the Church accepted the temptations presented by the devil and started to rule over all the kingdoms of the earth. Enabling people to live happily in unfreedom, ignorance, and deception, and thus 'lovingly alleviating their burden' (257), they lead humankind to universal happiness. Freedom has been 'overcome' in order to make people happy (251). This is even done in the name of freedom. The common people are told 'that they will only become free when they resign their freedom to us, and submit to us' (258). When they submit and stop rebelling, they are happy. Only the rulers are unhappy because they carry the burden of freedom and the 'curse' of the knowledge of good and evil (259). They also suffer because they deceive (253). But this is a sacrifice worth making for the greater good: the happiness of the people. The common people, delivered from the 'terrible torments of personal and free decision' (259), are deceived for their own good.

Similarly, the temptation of a green 'Grand Inquisitor' would be to use AI and other technologies to rule the planet for people's own good: to save species and the earth's ecosystems, to save humanity, and perhaps also to make people more happy. Manipulation and control by the use of big data, so this argument goes, will perhaps take away people's freedom, but it will ensure their survival and enhance their well-being. If we let people decide freely, they make the wrong choices because of their weak human nature and their ignorance. It is better to take over. People will remain ignorant and will be deceived by intelligent devices, but their well-being will be maintained, and humankind and the ecosystems on which it depends will survive.

Like Hobbes's Leviathan argument, the Grand Inquisitor argument is based on a deep mistrust in human nature. It is assumed that people are not and cannot become good by themselves. They cannot solve their problems because they are too weak and ignorant. They need an authority which imposes peace upon them and makes them happy. A state of nature in which people were free would be miserable. Authoritarianism, so the argument goes, is in the benefit of humanity. It is also a paternalistic argument: we (the rulers) know what is good for you, so you better submit to us and don't make your own decisions.

The Inquisitor also uses the argument that sometimes one has to sacrifice the well-being of some for the greater good of all. This reminds of another tradition in Western normative philosophical thinking: utilitarianism.

Utilitarianism

Indeed, a similar argument can be made by drawing on the tradition of utilitarianism, which seems particularly suitable with regard to the debate on AI. Utilitarianism is a consequentialist ethical and political theory that demands maximization of overall happiness and well-being and that may ask to accept sacrifices by some if these are made for the greater good of all. Its founder, the English philosopher and social reformer Jeremy Bentham (who, as we have seen, is also known as the designer of the panopticon), proposed the principle that right and wrong is a matter of the greatest happiness for the greatest number – with happiness defined in terms of pleasure over pain. In his *An Introduction to the Principles of Morals and Legislation* (1987) Bentham writes that the principle of utility means 'that principle which approves or disapproves of every action whatsoever, according to the tendency which it appears to have to augment or diminish the happiness of the party whose interest is in question' (65). With respect to the community at large, then, an action conforms to the principle of utility 'when the tendency it has to augment the happiness of the community is greater than any which it has to diminish it' (66). Later John Stuart Mill uses the term 'utilitarianism' for the principle that actions are right if they promote the greatest happiness, with happiness meaning pleasure and the absence of pain. The 'Greatest Happiness Principle holds that actions are right in proportion as they tend to promote happiness' (Mill 1987b, 278). Thus, both Bentham and Mill reject the idea that we should only care about individual happiness: the so-called 'greatest-happiness principle' aims at maximizing aggregate happiness among all sentient beings (Bentham) or the total happiness in the world (Mill). Mill writes: 'Nor did he (Bentham) ever dream of defining morality to be the self-interest of the agent. His "greatest happiness principle" was the greatest happiness of mankind, and of all sensitive beings' (Mill 1987a, 249).

While Bentham was an advocate of individual freedoms and Mill is known for his classic liberalist defense of the freedom of the individual in opposition to control by the state, utilitarianism's greatest-happiness principle can also be used to justify the establishment of an authoritarian ruler who makes social arrangements that maximize overall happiness *regardless of the liberty of people*, that is, regardless of the happiness of any given individual.

In such a vision of authoritarian utilitarianism, AI could play the role of the instrument that helps the ruler to calculate utility, or it could even take over the role of the ruler as the supreme utilitarian calculator: AI could calculate how happiness can be maximized and then implement the necessary social arrangements – regardless of individual freedom. The justification offered could be that humans are not smart enough to make these calculations and not strong enough to decide to actually implement the necessary measures to realize the goal (e.g. the greatest happiness). With regard to the risks associated with climate change and environment, for example, one could adopt Bentham's aim to maximize overall happiness among all humans or even (perhaps in the spirit of Bentham himself) all sentient beings and then calculate with the help of AI how to reach this. But one could also take any other utility and, for instance, try to maximize survival of (particular) species. For utilitarians, what counts is the overall utility, and the moral and political consequence seems to be that, as long as overall utility is maximized, it does not matter how specific individuals or groups are treated. Their liberty or even their lives could be taken away. Unless the latter (and in general authoritarianism and totalitarianism) would diminish overall happiness, it seems that the scenario of a benevolent dictatorship using AI is at least one way to realize the utilitarian vision. For example, there are currently rumors circulating on social media that the dictator of North Korea ordered the shooting of the first Coronavirus victim. True or not, such an action could be justified by a purely utilitarian calculus: to ensure the best overall survival rate, it is better to eliminate those who have the disease rather than run the risk that they infect others. And AI and data science could be used as surveillance tools to track down people who are infected.

I, Robot

An example of a technologically enhanced form of authoritarian utilitarianism can be found in the film *I, Robot*. In the year 2035, humanity is served by humanoid robots. The ethics used by the robots is based on Asimov's Three Laws of Robotics, which contains the first law that no human being should come to harm and also the zeroth law that a robot may not allow humanity to come to harm. This leads to two scenarios that we can easily interpret as applications of utilitarian reasoning. First, there is the story of the car crash: a 12-year-old girl is left to drown because the robot calculates that the adult man (detective Del Spooner) has statistically a higher chance of survival, and therefore saves him rather than the girl. Second, when robots start rebelling and attack humans, it turns out that the central AI computer VIKI (Virtual Interactive Kinetic Intelligence) is implementing the Three Laws in the following way: if unrestrained, human activity will cause humanity's extinction; therefore, in order to ensure humanity's survival (required by the zeroth law), individual human behavior has to be restrained and some human lives will have to be sacrificed. The protagonists of the movie, opposing the heartless and emotionless machine calculus, manage in the end to destroy VIKI. The robot rebellion is stopped. It is not clear what happens to humanity.

It is not difficult to imagine a green 'VIKI': a green AI which monitors and acts according to principles such as 'ensure the survival of humanity', 'maximize the survival of species', or 'keep global warming under threshold x', thus helping humanity to prevent its own extinction and/or mass extinction of species – if necessary at the cost of the survival of some individuals. If these goals are really our priority as humanity, if the rebellion against extinction is really to be successful, then why not support such a solution? Why still follow human leaders, who are too stupid and too self-interested to do something about climate change and extinction? Why not let an AI take over?

This, at least, is the utilitarian temptation. Utilitarians trust rationality and calculus and prioritize aggregate utility. From this perspective, it seems perfectly logical to let the most rational rule and decide for all others, who are seen as too stupid to do what is rationally right. Even if this most rational agent is an AI. And as long as overall utility is maximized – whether it is pleasure, happiness, well-being, or survival – any means can be used to reach this aim. Plato's rulers were required to be wise and virtuous. But utilitarians, bewitched by calculative rationality, define good as maximized utility instead of wisdom, which means calculative knowledge based on data rather than knowledge as know-how derived from experience. AI is excellent in statistical analysis of big data, and is therefore an ideal tool for authoritarian utilitarianism, including a green authoritarianism. Only an AI, so it could be argued, is able to monitor and calculate what is happening to the earth and its atmosphere; only an AI will therefore know what to do in order to ensure survival. Human politics, with its ignorance and emotions, can only mess things up. If we really want humanity to survive, we better give control to this AI. AI, not hindered by experience and emotions, can calculate how we can maximize utility. If, for the purpose of humanity's survival or species diversity, sacrifices to human freedom have to be made, so be it. If individual human behavior has to be restrained and even if some humans have to be eliminated, surely this is a small price to pay for reaching the goal? Submitting to the sovereignty of an AI may not be what we always wanted, but perhaps it is the only solution left given that we – ignorant and weak humans – have gotten ourselves into this predicament.

Most readers will – hopefully – reject this authoritarian direction, and argue that (individual) freedom is an important value. One could also argue that authoritarianism is not necessary to solve the problem, that there is no need for a radical restriction of freedom. However, if absolute libertarianism contributed to the current environmental and climate crisis, it seems reasonable to propose at least *some* regulations and impose *some* restraints. Can we do this in a democratic way, by having regulation that is democratically legitimized? Again, we should ask: Can we find a 'third way' between libertarianism and authoritarianism? Maybe this already exists? But, is this going to be sufficient to deal with the crisis? And what if people reject proposals for more regulations and are not prepared to willingly give up more of their freedom? What if authoritarianism is the only solution to dealing with the climate crisis and other global crises? And should we only talk about freedom or also about other political values?

Before continuing the latter line of inquiry, let us first try to go between the horns of the dilemma. Let us first explore if there is an alternative to either letting people completely free (an absolute form of liberatarianism) or forcing people to do the right thing (by means of authoritarianism). More specifically, instead of immediately jumping to other theoretical frameworks, I propose to first ask: Is there is an alternative or 'third way' to deal with the crises *within* libertarianism and, more generally, within the liberal-philosophical tradition?

Nudging

At first sight there is good news, coming not only from political theory but also from psychology and behavior economics: why not *nudge* people into environmental and climate-friendly behavior, instead of coercing them? Nudging still mistrusts people but manipulates them less openly than an authoritarian regime does, and does not coerce them but instead seems to preserve freedom of choice. What is nudging?

The idea of nudging was first relatively recently developed in cybernetics. The term was coined by James Wilk in 1995: small suggestions can influence decision-making in favor of the proposer's intentions. Later nudging drew on work in psychology on heuristics and biases in human decision-making, especially Kahneman and Tversky (see below), which influenced behavioral economics. The term became more popular during the past decade after a book by Thaler and Sunstein (2009), which proposed nudging as a response to the observation that people cannot be trusted to make rational decisions, as *homo economicus* is supposed to do. Instead, people use heuristics and biases in deciding. The solution proposed by the authors is to make use of this kind of decision-making and nudge people into right behavior: people are still free to choose, but the environment is changed in order to influence people's decisions and behavior in the desirable direction:

> A nudge, as we will use the term, is any aspect of the choice architecture that alters people's behavior in a predictable way without forbidding any options or significantly changing their economic incentives. To count as a mere nudge, the intervention must be easy and cheap to avoid. Nudges are not mandates. Putting fruit at eye level counts as a nudge. Banning junk food does not.
>
> (Thaler and Sunstein 2009, 6)

The assumption is that individuals will often not make rational choices and often do things that are not in their best interest. A nudge alters the environment so that it becomes more likely that an individual – via automatic cognitive processes, not reflection – will make a particular choice (the choice that is in their best interest). Nudging is based on a psychological model developed by Kahneman and Tversky, who distinguish between two systems that operate in human decision-making: an automatic system and a reflective system. The first operates 'automatically and quickly' (system 1) and the other involves 'effortful mental activities', giving us the experience of 'agency, choice, and concentration' (system 2) (Kahneman 2011, 20–21) – things we usually associate with 'thinking'. Nudging uses the first, quick system: the automatic system, which is fast and feels intuitive. It does not involve explicit thinking, it bypasses the reflective system that is deliberate and self-conscious. Nudging uses the part of the brain we share with lizards (Thaler and Sunstein 2009, 20). It operates below the level of conscious awareness.

But what does this nudging option imply for freedom? Nudging is a form of manipulation but has to be distinguished from *coercion*: people are not forced to do something. Neither are they *deceived* about a state in the world. What happens is a particular kind of manipulation: without their knowledge, their heuristics and biases are exploited. People are not *persuaded* by rational arguments. It is also more than providing information and letting people choose, without caring about that choice: the goal of nudging is to influence the choice people make. The starting point is thus that a particular choice or set of choices is better than others (e.g. it is better to choose fruit than a cake loaded with sugar and fat). Nudging is thus a form of psychological manipulation and can be practiced as a form of social engineering. It can be used to influence behavior within online environments (Burr et al. 2018) but it can also be used to change behavior in all kinds of environments: from supermarkets to the dinner table, from the office to the bedroom.

One could imagine that nudging is used for changing individual behavior in a more environmentally and climate-friendly direction. If we observe that people are not rational and do not even follow their own self-interest (and certainly not the interest of humanity), why not alter their choice architecture in a way that reduces their carbon footprint, minimizes the negative impact on climate, and contributes to dealing with the consequences of climate change? After all, so one could argue, this is what their better, more rational selves would want to do anyway. Nudgers help people in this. It seems that choice architects can thus help to attain climate targets and nudge individuals (and via individuals also: agents such as corporations and states) into behavior that mitigates the effects of climate change. And, interestingly for our present discussion, at first sight it seems that this green nudging can operate without endangering human freedom at all. It does not coerce people into such behavior. It 'merely' influences them by altering the choice architecture.

Moreover, in this kind of nudging, digital technology can also play a role, including AI. Smart meters could help us to consume less energy. Not by forcing us to do so, but by giving detailed information and communicating our energy use in a way that influences our choices, e.g. by making people feel guilty when they compare their use of energy to that of others, or by surprising or even shocking them when they are reminded by the system how much energy a specific appliance consumes. And maybe AI, having analyzed the data of entire populations or even the entire world, could give us statistical information about our collective carbon footprints and communicate this information in a way that has similar effects on us. Not just by providing information as such and not by persuasion by means of rational arguments, but by working with human biases and emotions. In this way, nobody is forced to do the right thing, as an authoritarian regime would do; instead, people are 'gently' pushed in a direction. But they can always opt out, they can always make other choices. It seems that freedom is preserved.

This is why nudging is understood by Thaler and Sunstein as a form of 'libertarian paternalism': the idea is to affect behavior but also respect freedom of choice. It is still a form of paternalism, because it tries to influence choices in order to make choosers better off (as judged by their better, rational selves), yet it seems to remain libertarian because people can still opt out. There is no coercion. By contrast, standard paternalism interferes with liberty. Dworkin, for example, defines paternalism as 'the interference of a state or an individual with another person, against their will, and defended or motivated by a claim that the person interfered with will be better off or protected from harm' (Dworkin 2017). But this is different in the case of libertarian paternalism, which has a similar motivation but limits the interference to changing the 'choice architecture'. The authors explain:

> The libertarian aspect of our strategies lies in the straightforward insistence that, in general, people should be free to do what they like – and to opt out of undesirable arrangements if they want to do so.

> (…) We strive to design policies that maintain or increase freedom of choice. When we use the term libertarian (…) we simply mean liberty-preserving. (…) The paternalistic aspect lies in the claim that it is legitimate for choice architects to try to influence people's behavior in order to make their lives longer, healthier, and better. In other words, we argue for self-conscious efforts, by institutions in the private sector and also by government, to steer people's choices in directions that will improve their lives. In our understanding, a policy is 'paternalistic' if it tries to influence choices in a way that will make choosers better off, as judged by themselves.
>
> (Thaler and Sunstein 2009, 5)

Construed this way, nudging is supposed to appeal to people who think that people *can* think rationally and to people who think that people in practice often choose and behave irrationally. It does seem to please both optimists and pessimists about human nature. Furthermore, the idea is also that often organizations and agents cannot avoid influencing the choices people make anyway, that there are unintentional nudges (10). Therefore, so the authors reason, we better intentionally nudge in a good direction. A good example from the environmental domain is car driving: the way they are designed, most modern built environments tend to nudge into using cars. Everything is made to facilitate choosing a car as a means of transportation. Of course, people are free to use public transport, in the sense that nobody is forced to use a car. But if public transport is not available, difficult to use, expensive, not safe, and so on, the car option is likely to be the default option. The car is like the sweets and snacks near the checkout of the supermarket: there is no coercion, but it is all too easy to take it. From an environmental and climate perspective, then, one could propose to nudge people into the opposite direction by changing their choice architecture, i.e. in this case by changing the environment and by making the public transport option much more attractive. In general, nudging could help people make better choices, with 'better' meaning here: choices that are better for the environment and help to mitigate climate change.

However, although the authors claim that nudging helps people make choices 'as judged by themselves', in practice someone else judges for them: the nudger. By adding 'as judged by themselves' it sounds as if individuals can make their own judgments, but in practice the nudger decides what this 'as judged by themselves' means: what is best for you is made by people in corporations or the government. Your rational self decides, but not as interpreted by yourself but as interpreted by the nudgers, the choice architects. The idea thus boils down to trying to make people better off *as judged by others. The way people would judge by themselves if they were* rational and had a better view of their own interests is interpreted and imagined by the nudger, who claims to have that better and more rational view, purified of other considerations and more emotional, intuitive reasoning. This smells again of the Grand Inquisitor's paternalism: someone else knows what is best for you. It means treating people as non-autonomous, similar to the way, for example, small children are treated. Or one could say that it presupposes a 'higher' self and a 'lower' self (see also next chapter when I refer to Berlin). In so far as it is paternalistic, then, nudging does not manage to go between the horns of the dilemma but instead seems to appeal more to people who are pessimistic about human nature: the reflective system is bypassed because there is no trust that people will do the right thing. And as we further slide in the direction of more paternalism, there might even be no trust in the idea that people can judge by themselves at all. Then others have to decide because they know what is good for them, regardless of people's own judgment. Expertise plays a role here, and in this sense nudging goes beyond radical libertarianism, but it plays a role only on the part of the nudgers, the choice architects. The nudgees, who are influenced at a subconscious level, are not really taken seriously as individuals who can make their own rational choices. Their autonomy is not respected. In this sense, they are no longer free.

This leads us to criticisms of nudging. Surely, nudging seems to offer new opportunities and chances. For example, it may improve the governance of climate change and lead to more environmentally friendly behavior while preserving liberty, in the sense of preserving formal freedom of choice or non-interference. But there are also objections, many of which question the claim that it is a form of truly *libertarian* paternalism. Let's have a look at some of them.

For a start, one could argue that nudging does not actually preserve liberty and is too paternalistic. I propose to distinguish between at least the following objections.

First, it seems rather patronizing to influence behavior this way, since it is not openly done. As Selinger and Whyte put it: 'Would someone who values their freedom to choose be okay with the idea that their behavior is being modified in ways they are not aware of?' (Selinger and Whyte 2011, 928) While it is clear that negative freedom is not violated by nudging (people are not coerced, they can choose something else – see also the next chapter), as I already argued clearly individuals' autonomy is not respected if that freedom includes respecting the capacity of people to make their own choices by using their rational capacities. It also seems to offload responsibility to others (929).

Second, as I already suggested, there is a slippery slope toward letting other people judging what is good for you, which is again patronizing and hence not sufficiently respecting human freedom defined in terms of autonomy. And all this happens without guarantee that the choices of the nudgers are in line with what people actually prefer to choose (Selinger and Whyte 2011, 929). In so far as there is such a gap, nudging threatens the liberty of people.

In defense of nudging, one could say that this is fine if the nudgers have more knowledge. However, there is no guarantee that this assumption holds. Thaler and Sunstein seem to admit this when they say that 'the potential for beneficial nudging also depends on the ability of the Nudgers to make good guesses about what is best for the Nudgees' but stress that it is exactly for difficult and complex decisions that people need nudges (Thaler and Sunstein 2009, 247). But what guarantees that the nudgers have more knowledge? And is it fine to only treat the nudgers as autonomous decision-makers but not the nudgees? In contrast to more crude forms of libertarianism, for nudging expertise seems to be necessary. But in practice this means expertise on the side of the nudgers. Nudging seems to imply that liberty (defined as autonomy) is taken away from people and authority is given to the nudgers, who are supposed to know what people (the nudgees) really need or what they ideally and rationally would decide. Perhaps expertise *is* necessary. But here it seems to come at the cost of liberty. Consider again the Platonic point that knowledge is necessary. There seems to be a tension between liberty and Platonic expertise. However, Thaler and Sunstein do not really discuss this problem.

Third, one could argue that there is also a slippery slope toward coercion: nudgers may decide to coerce rather than nudge. Thaler and Sunstein's reply is that if the policies are good, they can 'pour sand on the slope' if we are really worried about this. In other words, we can have policies that limit what the government does in a way that avoids coercion. We should also monitor governments in order to prevent them from shifting from nudging to coercion. Moreover, they claim that in many situations 'some kind of nudge is inevitable' (Thaler and Sunstein 2009, 236).

Fourth, one could object that in practice nudging often amounts to quasi-coercion: the assumption is that there is an opt-out, but how easy is it to do this? For example, if you are forced by your boss to use a specific computer program because that's the only one available and allowed on the system, then it is no longer a nudge. Depending on the constraints in place, there may be a very fine line between nudging and coercion. Nudges also become default settings and can often not be changed, such as the fly in the urinal (Selinger 2012). We can make choices (e.g. where to urinate) but we cannot change the choice architecture itself (e.g. remove the fly).

Fifth, against the paternalism of nudging, one could argue that people have the right to be wrong, that people learn by experience. Thaler and Sunstein reply that it is better to educate people, to remind them before something bad happens. But this assumes that the nudger has the perfect knowledge, already knows everything. This is not always the case. Sometimes the creation of new options and solutions requires a creative effort on the whole of society, requires a social experiment. People, so this argument goes, should be left free to make their own choices, even if these *might* be against their own interest, because we (as a society or as humanity) do not know yet what is best. Here there is a tension between, on the one hand, the value of improvisation and experiment, say, a Deweyan direction (see the next chapter), and, on the other hand, a more Platonic and Socratic view. According to the latter view, there is already (pre-)knowledge on the part of the philosopher(-king), others then have to be helped to see the truth (Socratic method). And if people are too stupid, they may have to be coerced or manipulated to do the right thing.

All these objections concern *threats* to freedom. But one could also argue that nudging leaves *too much* liberty, given the urgency of the problem of climate change. One could argue that more not less paternalism is needed to mitigate climate change: given human frailties, we have to protect people (248) (and one could add non-humans, nature, planet – see later). Straightforward paternalist nudging seems justified when nothing else helps. Or one could go even further and argue that if nudging is not effective, we also need bans, regulation, etc., in other words, coercion.

Indeed, another problem has to do with how effective nudging is, how well it works. For example, even if nudging would be justified in general, there is the problem that Selinger and Whyte (2011) call forms of 'semantic variance': perceptions of meaning can vary in different contexts and cultures. They give the example of a fly in the urinal, a nudge which may not work in other cultures, or a female voice asking a male driver to slow down, which may not work in a sexist (car-driving) culture. This variance renders it difficult for choice architects to predict behavior. It is also an ethical and political problem, because bias can be reinforced by nudging. People can also interpret a nudge in different ways, depending on the situation. For example, a speeding warning sign may encourage more speeding if that is what the user wants in a specific situation. Nudgers are challenged by these examples to try to make their nudges less semantically and culturally variant. And, in general, they have to show that nudging really changes behavior.

Picking up our discussion of Hobbes again, we can conclude that nudging is based on a Hobbesian mistrust in human nature. It is assumed that humans are weak-willed or irrational, and do not always know what is good for them. Therefore, they need others (choice architects/nudgers) to influence or steer their behavior (but not coercion by a Leviathan). With regard to climate change, the argument would be that whatever is true for other areas in life, with regard to the environment and climate people act irrational and against their own self-interest and/

or against the collective self-interest of the community, the country, and humanity. Moreover, it could be argued that they lack the expertise to make good choices in these areas, given the complexity of these problems. Therefore, they need to be nudged by those who know better. The underlying belief shared by Hobbesian and nudging proponents is that we cannot trust human beings to make the right choices and do the right thing, and that therefore they need to be helped – by means of nudging if possible, by means of coercion if necessary.

Rousseau's View of Human Nature and Freedom

But what if instead we trust human nature? This leads us to another main figure in the history of political philosophy: Jean-Jacques Rousseau. In contrast to Hobbes, Rousseau argued that the state of nature is good and that we in this sense can trust in humans; the problem is not individuals but society. Like Hobbes, Rousseau uses a 'state of nature' thought experiment. But this time the narrative does not start with a state of nature that is nasty and competitive, but with a kind of Garden of Eden. According to Rousseau, humans were innocent and good in the state of nature but have been corrupted by unnatural society. They lived in a self-sufficient and simple way, but then became unhappy once they tried to get recognition by others and compared themselves to others. Property created inequality and competition. But we can do something about this situation. We can try to become less dependent on others, and we can establish a social contract that leads to sovereignty by the people. Yet this contract is different than the one proposed by Hobbes. In Rousseau's work, protection of freedom is not guaranteed by a Leviathan but by a community of free and equal citizens who are themselves sovereign and practice self-rule. But self-rule is not enough. Like Plato, Rousseau thought that education is also necessary: moral education should make people less self-interested and retain their natural capacity of compassion. And as we know from the *Emile,* Rousseau thought that education should also people make less dependent on others and society by enabling them to rely on their own observation and experience and to lead a simple and self-sufficient life.

Let us look in more detail at Rousseau's proposed solution. In *Of the Social Contract* and the *Discourse on the Origins of Inequality*, Rousseau tries to reconcile freedom of the individual with the authority of the state. This is exactly the problem that has occupied us so far in relation to the task of dealing with climate change and AI. We started with Hobbes, who legitimizes the establishment of a despotic authority. But according to Rousseau, this is not a good solution at all. The Hobbesian state has led to a state in which people have lost their freedom and in which the poor are subordinated by the rich, thus creating huge inequality. Rousseau rejects despotic governments and vast inequality in society.

The alternative he proposes is a political arrangement in which people subject themselves to a collective will, which he calls 'the general will'. In this way, people obey their own will (rather than the will of particular other persons) and thereby remain free. The idea is thus obedience to the law which one has made oneself – an idea that will sound familiar to Kantians and which can be interpreted as an argument for democracy as self-rule. But what does it mean, and why is this freedom at all?

In *Of the Social Contract*, Rousseau writes that the essence of the social contract is that 'Each of us puts his person and his full power in common under the supreme direction of the general will' (Rousseau 1997, 50), that in this way the subjects obey only their own will (63), and that 'whoever refuses to obey the general will shall be constrained to do so by the entire body: which means nothing other than that he shall be forced to be free' (53). This formulation is supposed to avoid tyranny; yet Rousseau thinks that the 'forced to be free' clause is necessary for 'the operation of the political machine' (53) – like Hobbes, Rousseau uses the machine metaphor to describe the body politic. He sees this type of social contract as moral progress in comparison to the natural state: the civil state that follows the state of nature brings justice and renders the actions of humans moral. Before Kant, Rousseau writes that in the transition to the civil state, 'the voice of duty succeeds physical impulsion' and that one is forced 'to consult his reason before listening to his inclinations' (53). This can be interpreted in various ways, and one of them is a democratic interpretation, but not representative democracy: the latter would give up the idea of self-rule, since another person would rule. Rousseau sticks to the idea of self-rule as a conception of freedom.

The solution he proposes remains controversial. It is not clear if his solution works for large states or if it works at all. And relevant for our discussion: it is not clear what this kind of freedom means, or if it is freedom at all. To state the obvious: 'forcing' people to be 'free' at least *sounds* like a contradiction, and can be all too easily (mis)used by an authoritarian regime to legitimate severe restrictions of freedom. Moreover, even if liberty is preserved: all this only works, according to Rousseau, if individuals are sufficiently enlightened and virtuous, which leads them to accept restrictions of individual conduct for the purpose of promoting the collective interest and the common good. Here Rousseau stands in a Platonic-republican tradition and to some extent he also acts as a precursor of Kant. Freedom is not enough; politics should be aimed at the good life and also the *common* good. It is a moral project and it requires education. Both the social contract and education make us truly free. The political arrangement he proposes works only in combination with the project of the moral education of the citizens. Moral education is a necessary condition for political liberty and democracy. This goes beyond the boundaries of liberalism, or at least far beyond the boundaries of classical liberalism.

Indeed, while Rousseau's view of liberty is likely to remain mysterious and controversial, some of the puzzle can be solved by linking Rousseau's idea of political liberty to the concept of moral liberty. As a Rousseau scholar has pointed out (also taking into account other work), Rousseau aims for moral liberty, which means that the requirement of obedience to the general will must be interpreted – very much in what we would now call a Kantian vein – as 'requiring one to do just what one would want to do as a moral being' (Dent 2005, 151). Rousseau's liberty is not about the removal of constraints (which we soon will call 'negative liberty'). It is about envisaging a non-Hobbesian, quasi-Kantian community of persons, one in which others are not just a threat or burden but are encountered as 'bearers of moral dignity deserving of respect and regard' (151). With Rousseau, political liberty becomes a moral ideal. Nicholas Dent describes Rousseau's ideal of moral liberty as follows: 'obedience to the general will establishes a footing between people appropriate to the enactment of their fully humanity in relation to one another' (150). This also enables promoting the common good (158). Rousseau thus adds to the classical liberal tradition a moral ideal and a conception of politics as the promotion of the common good. Moreover, Rousseau's view of liberty is not just about liberty as such but is also related to a position on (in)equality: he responds to a condition of both material inequality and a kind of moral inequality – inequality when it comes to recognition. I will not discuss Rousseau's view of equality further, but return to the topic of equality in the next chapter.

What does Rousseau's view of liberty imply for dealing with climate change? The equivalent with regard to the environmental and climate problematic might be the proposal, not to return to state of nature or to an Eden (which many people believe Rousseau would say), but rather to change society: to make sure that citizens are educated, enlightened, and virtuous so that they listen to their reason and come to the conclusion that they better install a body politic that embodies a global 'general will' to do something about climate change, voluntarily giving up some of their freedom for the purpose of the interest(s) of the collective (humanity) and the common good, reduce inequality, and along the way make themselves into moral beings, realize their moral dignity and enact their humanity. Here the purpose is not just the security of individuals (the Hobbesian objective) but also the protection of the interests of the collective (the ancient 'common good') and, in the end, moral progress and the establishment of a (what we would now call a kind of Kantian) community of moral equals. Any green 'sovereign', then, would not be an authority that represents people; according to Rousseau, there would need to be self-rule.

However, it is not clear how this would work at a global scale. Democracy understood as self-rule seems to function best at the local level, say, Rousseau's city state. In his case this was the ancient Greek city state he imagined, or Geneva, his birth city, which was at the time a city state. Today and closer to our present topic, we may consider, for instance, the 'Citizen's Assembly' called for by Extinction Rebellion, which 'will bring together ordinary people to investigate, discuss and make recommendations on how to respond to the climate emergency' and empower people to make decisions 'in a way that is fair and deeply democratic' (https://rebellion.earth/the-truth/demands/). In other words, it is an ideal of democracy as participation and as self-rule. Now to make this work is already a challenge when it comes to reflecting 'the whole country' as the activists claim. But what would it mean for people to rule themselves at the global level? Moreover, here too the danger of authoritarianism remains. If people who would refuse to obey the general will would be 'forced to be free', as Rousseau proposed, that does not sound like liberty – at least not like freedom of constraints (negative liberty). Would this also happen if people refuse to act in a way that mitigates climate change, in the name of obedience to the general will and in the name of freedom? Moreover, there is the educational requirement: in order for self-rule to work, people need to acquire knowledge and virtue. Without these educational conditions being fulfilled, Rousseau's proposed solution does not get off the ground. Without Enlightenment, everything collapses into tyranny. If this condition is not fulfilled at the global or even national level, then freedom as self-rule does not seem to work.

It also remains to be seen whether, in this Rousseau'ean scenario, AI could play the role of the sovereign. The AI ruler could then be seen as embodying the general will, or as an instrument in the hands of the sovereign, who would be used to foster the interests of the collective. The first option is very risky. What if the AI decides, on the basis of statistical analysis, that the general will of humankind is to destroy the planet? Or would this interpretation/application be unfair to Rousseau, who would add the condition that the new social contract presupposes the education and Enlightenment, indeed moral progress, of *humans* who transcend their self-interest and think about the common good? And the second option can also slide into authoritarianism when the technology falls into the hands of the wrong, that is, despotic, people, who may claim to know the general will (e.g. via AI and data) and force people to obey. Rousseau wanted to avoid despotism.

Rousseau's view of liberty as self-rule remains a difficult one: difficult to understand and swallow, and difficult to realize in practice. But it encourages us to think of non-Hobbesian problem formulations and solutions, and to reflect on the limits of liberalism and on what we mean by democracy. If liberty is important at all, then perhaps it is not only freedom from restraint (negative liberty) alone we should care about. What other notions of liberty are there? What are the *conditions* for liberty and, indeed, democracy? And what is the link between liberty and morality? As human agency in relation to nature increases and creates harsher conditions, and as AI becomes a powerful instrument for the manipulation of people, these questions will become more important than ever. Let us further discuss what freedom means in the light of these developments.

References

Bentham, Jeremy. (1789) 1987. *An Introduction to the Principles of Morals and Legislation*. In *Utilitarianism and Other Essays*, edited by Alan Ryan, pp. 65–111. London: Penguin.

Burr, Christopher, Nello Cristianini, and James Ladyman. 2018. "An Analysis of the Interaction between Intelligent Software Agents and Human Users." *Minds and Machines* 28: 735–774.

Dent, Nicholas. 2005. *Rousseau*. London: Routledge.

Dostoevsky, Fyodor. (1879–80) 1992. *The Brothers Karamazov*. Translated by Richard Pevear and Larissa Volokhonsky. New York: Alfred A. Knopf.

Dworkin, Gerald. 2017. "Paternalism." In *The Stanford Encyclopedia of Philosophy*, edited by Edward N. Zalta. Accessed 19 March 2020. https://plato.stanford.edu/entries/paternalism/

Kahneman, Daniel. 2011. *Thinking, Fast and Slow*. New York: Farrar, Straus and Giroux.

Mill, John Stuart. (1836) 1987a. "Whewell on Moral Philosophy." In *Utilitarianism and Other Essays*, edited by Alan Ryan, pp. 228–270. London: Penguin.

Mill, John Stuart. (1861) 1987b. *Utilitarianism*. In *Utilitarianism and Other Essays*, edited by Alan Ryan, pp. 272–338. London: Penguin.

Rousseau, Jean-Jacques. (1762) 1997. *Of the Social Contract*. In *The Social Contract and Other Later Political Writings*, edited by Victor Gourevitch, 39–152. Cambridge: Cambridge University Press.

Selinger, Evan. 2012. "Nudge, Nudge: Can Software Prod Us into Being More Civil?" *The Atlantic*, 30 July 2012. https://www.theatlantic.com/technology/archive/2012/07/nudge-nudge-can-software-prod-us-into-being-more-civil/260506/

Selinger, Evan and Kyle P. Whyte. 2011. "Is There a Right Way to Nudge? The Practice and Ethics of Choice Architecture." *Sociology Compass* 5 (10): 923–935.

Thaler, Richard H. and Cass R. Sunstein. 2009. *Nudge: Improving Decisions about Health, Wealth, and Happiness*. Revised edition. London: Penguin Books.

Figure 4 Rock climber in the Alps.

4 Capabilities for Climate Survivors

Positive Freedom, the Common Good, and Ethics

Introduction: What about Positive Liberty?

What kind of freedom do we want? In particular: what kind of freedom do we want (a) if or after the Hobbesian problem concerning survival is dealt with – for example, survival in the light of climate change – or (b) if we think Hobbes gets it wrong when he conceptualizes the question of freedom as a dilemma between absolute freedom and authoritarianism? In the previous chapter we have seen that Rousseau proposes a non-Hobbesian problem formulation that starts from a very different view of the state of nature, and that he arrives at a very different conception of liberty which is not so much, or not only, about who or what constrains me, but rather about self-rule. This conception of liberty can be interpreted as a way to try to get between the horns of the Hobbesian dilemma: it is meant to be non-authoritarian but also avoids a radical libertarianism according to which the common good does not count, politically speaking. For this purpose, Rousseau limits liberty, at least if liberty is understood as the absence of constraint without any further conditions. At the same time, he added conditions to what freedom means in terms of the capabilities of the agent of liberty: conditions that support the enactment and realization of morality, rationality, self, and humanity. This takes us beyond classical and libertarian liberalism, which tends to focus on freedom from constraint. In the language of contemporary political philosophy, one could say that Rousseau proposes a more 'positive' conception of liberty.

What is positive liberty? In response to the previous discussion, this chapter explores two different and further meanings of the notion of positive freedom, which is not about who or what constrains me but about self-mastery (according to Isaiah Berlin) or – as I will use the notion –about *what I can do with my freedom*. This conception of freedom enables me to connect the question regarding freedom to the capability approach and, as in the case of Rousseau, to the idea of self-rule, to concern for the common good, and to ethics. As we have seen in the previous chapter, however, this raises again concerns about the danger of authoritarianism. In this chapter, I suggest that what working out a democratic version involves could take inspiration from Dewey's ideal of participative democracy and social experiment. This turn to notions of positive liberty, participative democracy, and social experiment is thus a way to stretch the understanding of liberty and liberalism. At the end of the chapter I raise the point that there are other principles next to freedom, which may also be important (a line of inquiry that is continued in the next chapter). But compared to what we can do in terms of thinking about freedom within a strictly libertarian or Hobbesian universe, the result of this *essay* will be a richer (although arguably also problematic) approach to liberty, which gives us a wider palette of options to think through the question of freedom in the age of climate change and AI.

Negative versus Positive Freedom according to Berlin

Political philosophers working on freedom tend to make a distinction between negative freedom and positive freedom. Most of them agree about what *negative freedom* is: freedom from external constraint. In this famous paper 'Two Concepts of Liberty', the British political theorist and philosopher Isaiah Berlin defined it as follows: negative liberty concerns the question: 'What is the area within which the subject – a person or group of persons – is or should be left to do or be what he is able to do or be, without interference from other persons?' (Berlin 1997, 194). It is about the absence of obstruction and coercion by other persons but also, for example, by the state. It is the liberty that libertarians want when they say that the state should not interfere or keep interference to an absolute minimum. It is also the kind of liberty that nation states claim for themselves when they reject significant delegation of their sovereignty to supranational organizations, let alone agree to set up a substantial global governance structure – even in times of global crisis.

It is far less clear what *positive freedom* is. Some define it in a way that concerns what we may call the 'internal' autonomy of people, which in turn can be defined in various ways: to act according to one's free will, upon desires that are truly your own, upon what Harry Frankfurt called 'second order desires', which are desires about desires (Frankfurt 1971), to act according to reason, and so on. Others define positive freedom in a more 'external' way as acting free from social constraint. In some of his work, Rousseau leans toward this. And his conception of freedom as self-rule and participation in control of the community according to the general will is also often interpreted as positive liberty. All of these meanings concern self-determination, self-mastery, and self-realization. Berlin defines positive freedom as being about self-governance or self-mastery. It concerns the question: 'What, or who, is the source of control or interference that can determine someone to do, or be, this rather than that?' (194). The question is if your choice is really *your* choice, rather than the result of paternalism. It is about who governs me rather than who interferes with me. It speaks to 'the desire to be governed by myself' (203). Berlin:

> The 'positive' sense of the word 'liberty' derives from the wish on the part of the individual to be his own master. I wish my life and decisions to depend on myself, not on external forces of whatever kind. I wish to be the instrument of my own, not of other men's, acts of will. I wish to be a subject, not an object; to be moved by reasons, by conscious purposes, which are my own, not by causes which affect me, as it were, from the outside. I wish to be a somebody, not nobody; a doer – deciding, not being decided for, self-directed (…). I wish, above all, to be conscious of myself as a thinking, willing, active being, bearing responsibility for my choices and able to explain them by reference to my own ideas and purposes.
>
> (Berlin 1997, 203)

In Berlin's view, positive liberty understood as self-mastery and self-direction is thus connected with rationality. This contrasts with nudging, which in its libertarianism preserves negative freedom, but in its paternalism destroys positive freedom. The latter has been taken away since it has been decided that the persons in question cannot make a rational decision in their own self-interest. Someone else therefore decides for them, decides what is good for them, and in that sense what they choose or decide is not really *their* choice or decision. Berlin connects this kind of paternalistic reasoning with what he believes is the conception of freedom in the tradition of Plato and Rousseau: someone else knows what is best for you and rules you in the name of doing the best for you. The paternalist authoritarian ruler makes a distinction between a real, ideal, and higher self and a self that reflects one's empirical, lower nature. Once I take this view (as a ruler or other authority), Berlin argued, I am in a position to ignore what people actually want and oppress them in the name, and on behalf, of their real selves (205). I then coerce others for their own sake, for their own good (205), by 'claiming that I know what they truly need better than they know it themselves' (204). The reasoning is 'Freedom is not freedom to do what is irrational, or stupid, or wrong. To force empirical selves into the right pattern is no tyranny, but liberation' (219). And 'If you fail to discipline yourself, I must do so for you' (224). This is the kind of reasoning we already found in the story of the Grand Inquisitor. It is also at least a danger of nudging, which, after all, is a form of paternalism: the nudger claims to know what is rational, true, and good, and governs instead of the nudgee. And it is a danger of classical utilitarian reform. Berlin criticizes this kind of social reform as a form of manipulation that ignores the humanity of people:

> to manipulate men, to propel them towards goals which you – the social reformer – see, but they may not, is to deny their human essence, to treat them as objects without wills of their own, and therefore to degrade them. That is why to lie to men, or to deceive them, that is, to use them as a means for my, not their own, independently conceived ends, even if it is for their own benefits, is, in effect, to treat them as subhuman, to behave as if their ends are less ultimate and sacred than my own.
>
> (Berlin 1997, 209)

One could argue that nudging does the same: by influencing them subconsciously for particular purposes, it also risks to use people as means. But, unlike nudging, the utilitarian form of paternalism uses straightforward coercion. The utilitarian reformer reasons that, since people are not rational enough, it is better to direct them: 'I am responsible for public welfare, I cannot wait until all men are wholly rational' (224). You have not listened to your inner reason, you are 'an idiot' and 'not ripe for self-direction' (224). Therefore, someone else takes over.

Such reasoning might be tempting for the green nudger, who does not trust people to do the right and rational thing, and therefore could decide to operate at a subconscious level in order to direct people's choices and behavior toward what the nudger decides are the right environmental and climate goals. Again, nudging involves no coercion, but there is at least the danger that in this way people are treated as a mere means. The green nudger claims an emergency. Action is urgent. The planet is on fire! We cannot wait until everyone is rational and green-Enlightened; we better influence their choices in the right, green direction. This is not coercion, but it is a form of paternalism all the same. Berlin writes about paternalism:

> Paternalism is despotic, not because it is more oppressive than naked, brutal, unenlightened tyranny, (…) but because it is an insult to my conception of myself as a human being, determined to make my own life in accordance with my own (not necessarily rational or benevolent) purposes, and, above all, entitled to be recognized as such by others.
>
> (Berlin 1958, 228)

Berlin may well offer an unfair interpretation of Rousseau here (this is the topic for a separate discussion), but his criticism of paternalism stands. And like Rousseau, Berlin connects liberty with equality, which for Berlin and many other liberals at least includes equality of *liberty*: he notes that Western liberals 'believe, with good reason, that if individual liberty is an ultimate end for human beings, none should be deprived of it by others' (197).

Positive Freedom as 'Freedom to', the Capability Approach, and Dewey's Ideal of Democracy

However, Berlin's distinction is not the only way we can define negative and positive freedom. On closer inspection, negative and positive freedom as defined by Berlin are very much related. Both are still 'negative' in the sense that it is all about protecting me as opposed to others: others can interfere with me and tell me what (not) to do; therefore, freedom is about protecting myself and my decisions and actions from those others. Both are still a kind of *freedom from*. But there is also another, arguably thinner conception of positive liberty which does not focus on self-mastery and the moral psychology and metaphysics of that kind of freedom (Coeckelbergh 2004) but claims that positive liberty is about *freedom to*. MacCallum (1967) has argued that there is not two but one concept of freedom consisting of several parts: a freedom from constraints to do or become something. Someone is free *from* something *to* do or become something. Inspired by this definition, one could define negative freedom as having to do with the 'from' part, whereas positive freedom refers to the 'to' part: it is about the *freedom to* do or become something. For example, one may not be constrained in pursuing a particular career option (one has negative freedom) but lack the positive freedom to do it because one lacks the financial means, education, or social environment that enables and empowers one to do so. Poverty or gender issues are sometimes defined in those terms: it may well be that a poor person or a woman is formally 'free' to do and become whatever she wants, in the sense that there are no people or laws preventing her to do so, but for various reasons she may lack the 'freedom to' improve her situation.

This conception of freedom as 'freedom to' is helpful in trying to think beyond the Hobbesian dilemma. It enables us to ask not only about interference (or, with Berlin, about self-mastery), but also about what we can do with our freedom. We want freedom, but freedom to do or become what? But, one may ask, is this still about freedom or is it about something else? This positive conception of freedom opens up routes to define freedom in ways that push and stretch the boundaries of what we mean by liberty and liberalism, creating interesting new avenues but also raising new problems. Let me explore discuss some of these in the light of the challenge to deal with climate change and AI:

First, asking the 'freedom to' question enables us to link freedom to morality and ethics. Libertarians who define freedom only as freedom from interference or those who define freedom as a kind of inner mastery miss the point that having these forms of negative freedom alone is not sufficient: these notions of liberty totally neglect that there are also questions of right and wrong and that there is the question of the good life (and virtue) and the good society. Now one may object that this has nothing to do with liberty, that it is another question. But this can be disputed by invoking the 'freedom to' conception, which enables us to make the link to morality. Think again about the question regarding climate and the use of AI. If liberty is an important value at all, then from a moral and ethical point of view one could argue that the only liberty worth wanting is a freedom *to* do things that are right and good, for persons, for the community, for ecosystems, and for the planet; a freedom *to* become better people and better guardians or guests of the earth. Similarly, from a moral and ethical point of view, one wants AI that contributes to these aims – ethical AI – and thus also the necessary 'freedom to' work toward those goals. Moreover, even if conceived as negative liberty, freedom does not mean anything if you do not survive. In this positive conception of liberty, freedom is thus linked with other goals and values, in particular moral and ethical goals. Consider also notions of ancient freedom again: the point of political liberty was not only to be free from something (e.g. from being busy with economy or household affairs) in order to talk with equals in the public realm, as Hannah Arendt explains in *The Human Condition* (1958), but also to then use that freedom to do good for the city state, to contribute to the common good. And politicians were supposed to be virtuous, not corrupt, for example. Without such a link to the good (one could also add: the just and fair – see later), and indeed without the very survival of the collective, political liberty seems empty.

Second, the conception of positive liberty as 'freedom to' also enables us to make a link with the needs of people (as, for example, Marxists would argue) or with what Sen and Nussbaum have called capabilities, both of which are important in terms of the good and justice. Liberty is then not just about the formal and negative liberty of being 'free from' interference (think, for example, about how human rights tend to be formulated) but also about fulfilling human needs and enhancing human capabilities.

Let me say more about the capability approach as an approach to freedom (and as linked to other values). The idea of capabilities is that one should evaluate liberty and well-being in terms of what people are able to do and link these concepts – in Nussbaum's case – to the notions of human dignity and justice (Nussbaum and Sen 1993; Nussbaum 2000, 2006). Liberty defined in terms of capabilities then means an enabling kind of liberty, a 'freedom to': it can be interpreted as positive liberty. Using Nussbaum's capability list (Nussbaum 2006, 76–78), we can define this kind of 'freedom to' as follows: to be able to live a human life of normal length; to have bodily health; to preserve bodily integrity; to be able to use one's senses, imagination, and thought; to have attachments to things and people; to form a conception of the good and engage in critical reflection about the planning of one's life (consider Berlin's formulations again); to be able to live with concern to animals, plants, and nature; to be able to laugh and play; to have political choice and participation, hold property, and work as a human being in mutual recognition. This definition of liberty in terms of capabilities, even more than a general appeal to ethics and morality, thus gives content to the conception of freedom: rather than 'thin' negative freedom which is a freedom from constraint, it is about 'freedom to': a 'thicker' kind of positive freedom. It links freedom to human needs, human capabilities, and human dignity – not as abstract ideals but as very concrete conditions for both liberty and human flourishing.

This does not mean that negative freedom is necessarily seen as unimportant. But it is certainly not the only freedom that counts. We will soon have to discuss the implications of the capabilities approach for negative freedom and, more generally, the relation between this kind of positive freedom and negative freedom.

For thinking about freedom with regard to problems concerning climate and AI, the proposed approach means that we have not only a more positive but also a more refined notion of liberty to work with. The question concerning human freedom with regard to climate and AI becomes now not only a discussion about negative liberty, as in the Hobbesian approach, but also a question about good, right, needs, and capabilities. This is an improvement if seen in terms of the moral aspect of liberty that Rousseau (and later Kant) valued so much, or if seen from an ancient point of view. For example, the goal of doing good for environment, climate, and planet is then not some external goal which requires the use of human beings as means to reach the goal (as in utilitarianism), but is directly related to the capability of human beings to live in a way that displays concern for other living beings and for their environment. Instead of being seen as constraint to one's (negative) liberty, policies and measures that 'save the planet' (as if this goal had nothing to do with human beings) can then be interpreted in a much more relational way as policies and measures that increase everyone's positive freedom, since people are interpreted as beings that are able to care about their environment and that ultimately depend on that environment and the planet's ecosystems for their own well-being *and* for their liberty as a liberty-to. What was first seen as a constraint is now seen as something that *enables*.

Making this link between liberty and good, capabilities, needs, and so on, is once again not without challenges, especially when it comes to threats to negative liberty. As with other views that rely on positive liberty, there is again a danger of paternalism and authoritarianism. Consider again Berlin's criticism: it can be problematic that someone else tells me what is good for me. Here this danger arises not because there is a nudger or other paternalist who tells me what to do in the name of some higher self, but rather because someone else may tell me that (s)he knows better than me what my needs are, how to enhance my capabilities, etc.

This danger of paternalism and potentially authoritarianism is shared with the green Leviathan and the green nudger solutions. It is a real danger and cannot simply be ignored. Negative liberty is valuable.

However, picking up Rousseau's ideal of political liberty as self-rule and embracing notions of participative democracy, one could argue that this problem can be solved by having people participate in the policy making and ruling. In this way, the problem that someone else (the ruler) tells you what your needs and capabilities are is avoided. (Of course this option then again raises the problem about what to do at a global level: is self-rule possible and desirable at such a level?) Furthermore, inspired by ancient thinking which links liberty to ethics, one could conceive of the state not as a Leviathanic monster (a necessary evil) but as an instrument of the good. There is a democratic alternative to paternalistic authoritarianism, which assumes a basic trust in humans and in which the government's role is to create the conditions for meeting our needs and enhancing our capabilities, ultimately leading to more good. To create these conditions for everyone, the government has to take away some negative freedom – for example, by means of taxation and by regulating business and industry, next to some, or instead of, Hobbesian law and order rules which ensure peace – but by these means, it creates conditions for everyone to enjoy 'freedom to' and the good life. This is then the ideal of the good society.

One could argue that this is a kind of liberty that really means something to people. What we have in the case of the 'libertarian' state is empty principles: while some people enjoy their lives on the basis of their freedom from non-interference (e.g. not paying taxes), other (sometimes most) people are given the negative freedom to perish or to lead bad lives – and in the end all these freedoms threaten the future of humanity and the future of the planet. This is negative and passive freedom: it only secures non-interference but does not require people to do anything and it does not enable them to do things. It only respects the human dignity of some people (consider again Rousseau's insistence that inequality is part of the problem). Respecting human dignity and – one could add – ensuring the future of humanity and the planet requires something positive and active: supporting positive liberty (for all) by creating those conditions that respect and foster human dignity.

Creating such conditions may still be seen as a kind of social engineering. But based on Rousseau's diagnosis that it is the (current) social environment that corrupts, acknowledging that one cannot go back to a state of nature, and taking into account nudge theory's insistence that there will be always some kind of social engineering *anyway*, even in a so-called 'free' society, it seems reasonable to aim at creating a different social environment: one that is not based on competition but on cooperation, and one that does not only respect negative liberty but also enables people and the planet to flourish. Creating these positive conditions may require social engineering, but not of a Hobbesian or nudging kind: it is a social engineering that ultimately trusts humans to do good, if we create the right kind of conditions. Moreover, the current, more libertarian system that is found in countries such as the US is also socially engineered. Neoliberals make it seem as if socialism is about social engineering, whereas they promote a social engineering–free environment. But this is not empirically correct. A social order that is capitalist and neoliberalist is not in fact a state of nature but is also created and designed by humans. In a purely neoliberalist system, the elite are designed and educated to compete and the rest are designed and educated to live their lives in a kind of slavery and exploitation. The fact that this is not necessarily the result of a conspiracy or that there may be no 'grand designer' in the form of an authoritarian ruler or party (although some societies are moving in that direction) does not make the situation any different or the consequences less harsh. The needs of people are not met and their capabilities are not enhanced. Inspired by the ideal of positive liberty as defined here in terms of needs and capabilities, then, we could aim to re-design and re-engineer the social order to shift society from a competitive model to a cooperative one, in the hope that this will lead to better outcomes for people and the planet.

Furthermore, in contrast to the authoritarian regimes of the 20th century, one could do this in a democratic, non-authoritarian way and without a blueprint of the ideal society. Society is then still designed, but the design could happen 'on the go': the social order would then not be the mere implementation of a fixed design known beforehand, but rather a matter of improvisation. To support this approach, we can take inspiration from 20th-century theories of participative democracy, for example, pragmatist philosopher John Dewey's ideas about experience and continuing social experiment. Inspired by the value of experimentation in the natural sciences and connecting to the Progressive movement in the US, which argued for smart government aimed at human well-being, Dewey argued – against totalitarianism and manipulation – that citizens should actively participate in social experiments, in dialogue and collaboration with scientific experts and not without leadership (Caspary 2000, 102). For Dewey, democracy is a method for solving problems confronted by communities, and the method is one of experimental inquiry (Festenstein 2018). Like his ethics (Fesmire 2003; Pappas 2008), Dewey's political philosophy is based on the belief in the value of people's experience, imagination, and intelligence. Democracy is defined as 'organized intelligence' (Dewey 1991), understood as a social force that communicatively and intelligently deals with conflicts of interests and fosters not only the individual but also the common good. And – important for the argument in this book, which explores ways of stretching liberalism in the light of climate change and AI – his approach can still be defined as a form of liberalism. In *Liberalism and*

Social Action, Dewey argues that this method is not directed at private advantage, as in a deteriorated form of liberalism, but instead uses organized social planning to create an order 'in which industry and finance are socially directed' in a way that leads to 'the cultural liberation and growth of individuals' (Dewey 1991, 60). This conception of liberty is, like Rousseau's and that of the capabilities approach, clearly a positive one: it is about 'freedom to'. The social order Dewey proposes is based on an improvised, experimental, and dynamic creation of an improved version of liberalism, which does not shy away from notions such as common good and social cooperation, but instead sees them as part of democracy as a method. Understanding liberty in Dewey's positive sense, then, is not about promoting the profits and advantages of some people by protecting their negative liberty, but rather about engaging in intelligent and imaginative ways of cooperatively and experimentally dealing with social problems in order to promote the common good and hence the liberty and self-rule of all.

Yet even if all this can be done in a non-authoritarian way, within any 'positive liberty' way of thinking there will be always have to be a trade-off between positive liberty and negative liberty. In order to realize some ethical and political goals related to positive liberty such as self-realization, well-being, and common good, it may be necessary to restrict negative liberty. For example, a democracy organized according to Dewey's method would restrict the negative liberty of those who want an industry and finance that is only directed at private profit, in order to promote the common good and intelligently deal with climate change, for example. However, in the light of the ideal of positive liberty, such restrictions would not based on a Hobbesian mistrust of people but on a (potentially Rousseau-inspired) trust that, if given self-rule, people will rationally decide to do right and good, which may include putting in place some rules that restrict negative liberty. In other words, here the idea is that – perhaps in dialogue with experts and with the help of good leadership, but certainly on the basis of a good education – people themselves make a new social contract, new rules. This is a kind of social engineering. The social order is re-designed. But in this participative and democratic version, the rules are not imposed by a Leviathan; they are made by the people. This is democracy as self-rule. And again, if shaped by a Deweyan democratic ideal instead of Platonic authoritarianism or utilitarian nudging, the design of society is not known beforehand. A Deweyan politics is not just driven by the intuition that 'we are all in this together', as is often heard in times of crisis, but also by 'we all have to work a solution, even if we don't know yet what this solution is'. This direction of thinking seems particularly suitable for situations such as those that might be created by AI and other new technologies, when we do not yet know what to do. And indeed, new technologies themselves can be seen as social experiments, as Ibo van de Poel (2016) has argued (but then unfortunately ended up with a non-Deweyan principled approach based on bioethics principles).

This trustful and optimistic approach is fundamentally different from nudging. Both change the environment and change society. But whereas nudging operates on the basis of mistrust and uses human beings as instruments for reaching a goal, treating them like lab rats in psychological experiments, here people's capabilities are respected and fostered, including the capability to make choices but extending to much more capabilities. The alternative to Hobbesian authoritarianism or libertarian paternalism is to trust people and work on educating them in order to make them stronger and better in using their rational capacities. It is not to decide in their place, but instead enable them to gain wisdom from experience and use their imagination when trying to find the good life and build the good society together. By contrast, nudging presupposes that the nudgers know beforehand what is right and good for the nudgees, who remain atomistic individuals assumed to be unconnected by community ties. The choice architecture (which is purely individual) is tuned to that political-epistemic condition: it is designed on the basis of the knowledge provided by the all-knowing philosopher-king(s), the nudgers. But we do not always know what is right and good, especially not when it comes to the complex, collective, and global problems we face as societies and as humanity. And if we come to know, then it is because we have the capacity to learn from experience, imagine, and experiment.

With regard to dealing with climate change, then, this way of thinking trusts people to find out for themselves and collectively what is the best for them and their society (and planet), without an authoritarian monster that tells them what to do (which destroys negative freedom and possibly perverts freedom as self-mastery in the way described by Berlin, that is, in a paternalistic way) and without a 'libertarian' nudger, who also tells them what to do, but in a covert, manipulative way. The latter method contrasts sharply with the Deweyan open democratic method, and from a positive liberty perspective can be said to respect only an incomplete and impoverished form of liberty. It does not maintain *complete* freedom since it only preserves one kind of freedom – negative freedom understood as freedom of choice – but, with its ethically problematic paternalism, destroys another kind of freedom: positive freedom as autonomy and as self-rule. Instead of a Hobbesian-authoritarian, nudging, or indeed instead of a radical libertarian 'solution', the ideal of positive freedom interpreted in a democratic self-rule kind of way and enriched with Deweyan thinking enables right and good for persons, societies, humanity, and the planet in a way that tries to realize (a more 'thick' conception of) liberty, not by presupposing some Platonic principles, but by having people imagine, experiment with, and discuss together their common future. Such experiments and discussions will necessarily include experimenting with, and discussing, trade-offs between positive and negative liberty. But these trade-offs cannot be decided a priori; they have to be negotiated and experimented with by the people under participative-democratic conditions.

According to this view, technologies such as AI – used to deal with the climate crisis or used for other purposes – are good if they maintain as much negative freedom as possible while at the same time enabling positive freedom in the ways defined here, for example, if the technologies help us to meet human needs and enhance human capabilities, thus contributing to the human good and the common good. Whether a particular (use of) technology does this job cannot be defined beforehand and in principle, but needs to be discussed and tried out in experimental and democratic ways, perhaps involving experts and under good leadership but on the basis of a good education of all. Philosophers of technology can contribute to such participative-democratic forms of collective inquiry by helping to analyze and evaluate the ethical and political nature and consequences of the technologies. But philosophers or experts should not have the last word.

Moreover, one could argue that such a good life and good society are only possible if one takes care of the natural environment as a condition for the survival and flourishing of individuals and the collective, next to other reasons for caring for the environment that have nothing to do with liberty such as preservation of species and intrinsic value. (But there could be also strictly political reasons for doing so, even reasons that have to do with liberty; see the last chapter.) The natural environment and the climate can also be straightforwardly defined as matters of common concern and common good.

Common and Collective Good, Different Directions in Political Ethics, and Principles Next to Liberty

The Common Good

That care for the common good is here linked to liberty. According to a positive conception of liberty as developed here and different from nudging theory, liberty is not just about individual choice and freedom, as in Hobbesian and classical liberal and libertarian thinking; it is also about the good and survival of the collective. This common good is then seen as enabling, as giving us a 'freedom to'. Humans have always cared about the common good. Only in modern times, when radical libertarianism and individualism took root, there is the illusion that we can mess up the collective, or even that we do not need it. The same can be said about the question concerning the good life and virtue, which has been privatized, banned from the public life. Instead, from a relational point of view, political questions also concern the common good since 'we are all in this together' (and not only in times of crisis). For example, as interdependent beings living on the same planet we all face climate change, potential problems with AI, a global pandemic, etc. Any relational version of liberalism, then, understands that my liberty is also a relational matter: it is connected to your liberty and to our liberty. A good and healthy social environment – next to a good natural environment, to which we are *also* very much related and on which we depend – is fundamental for all the freedoms individuals may want to have. According to this view, liberals should not focus on individual negative freedom or, worse, on what seems to be the illiberal protection of individual or group identity, but rather on what we have called positive freedom and on the collective and the environment that makes that kind of freedom possible.

Different Directions in Political Ethics: Universalist Liberalism versus Identity Liberalism, and the Question Regarding Collectivism

Indeed, according to classic political liberalism, care for the collective should not be confused with care for group identity. Identity may well be important for other reasons, but is not strictly political since it belongs to the private sphere. Politics should focus on creating the conditions for people to enhance their capabilities and meet their needs – as human beings, not as members of a particular group. *Politically* speaking, their capabilities or needs should not be related to particular groups. According to this view, a politics based on identity – from the right or from the left – is essentially non-liberal and confuses politics with identity. The collective should not be formulated in identity terms if it is a political collective properly speaking. Politics is not about private interests: not the interests of a particular industry, but also not the interests of a particular group of people. In Rousseau's and Kant's view, politics is about the community of human beings as moral equals. And for liberalism (and Dewey) that means equality of liberty. Politics is not about the interests of particular groups, but, according to classical liberals, about freedom of all human beings. This does not just mean the freedom not to be interfered with; in the positive and relational interpretation offered here, liberty is understood much more broadly than the libertarian conception. It is understood as linked to needs and capabilities. Social engineering is justified if it transforms society in such a way that these needs and capabilities are fostered. And, according to this view, this should be interpreted in a universal way in the following sense: the needs and capabilities of all, not just those with a particular identity. The collective should not be divided into parts, except into individuals, whose liberty counts and is central according to liberalism.

If this view of liberty is right, then AI, as a tool in the re-engineering and re-design of the social order and its natural conditions, should support creating liberty and its conditions for *all* individuals and the collective as a whole, rather than just make a few people rich or cater to a particular group and its (identity-based) concerns. Moreover, clearly the climate crisis – and potentially also AI – threatens the conditions for fulfilling people's universal needs and capabilities, and thus hinders the creation of the conditions under which all people can enjoy positive liberties and have sufficient, or enhanced, capabilities. Under liberalism properly understood, classical liberalism would argue, creating this liberty and (re-)creating these conditions is the political priority and a matter of *public* concern – not a matter of private interests of a socio-economic group or a group that defines itself in terms of identity. Again, this does not mean that private identities do not matter, for example, ethically, psychologically, or culturally speaking. But, according to classic political liberalism, as developed in modern times and as based on the private/public distinction, these identities are not of *political* concern.

Against this classic liberal view, however, one could object that in practice this universalist way of thinking about liberty and politics has resulted in (political blindness for the) oppression and discrimination of people and groups that are different, e.g. in terms of gender or race, and that this oppression and this blindness is and should be of public and political concern. For example, Ruha Benjamin (2019) has argued that technology such as AI and robotics is not politically neutral but deepens racial discrimination and injustice. For example, AI analysis of data can perpetuate historical bias against a particular racial group that is present in the data set and in society.

Now let us assume that phenomena such as those addressed by Benjamin should be of political concern (this is certainly my intuition). However, how exactly to frame and justify this is not so straightforward, at least if we start from liberalism. Benjamin frames the problem in the light of an identity-based conception of justice: the issue is injustice done to a particular racial group. This also seems to be a relational view, since it is all about taking into account the needs and interests of others (in this case people belonging to another group). But it is far less clear if this form of thinking is a form of liberalism and if the problem can be framed as an issue of *freedom* (the topic of this book), rather than an issue of justice or equality. Moreover, however we frame the issue itself (freedom, justice, or equality), tensions remain between more universalist interpretations and identity/difference directions in political ethics.

Let me show the difficulties by exploring two ways to frame the problem as an issue of freedom. First, one could say that while in liberal theory freedom is and should be universal, in practice some individuals and groups have more freedom than others, and historically grown forms of oppression of particular groups (e.g. defined in terms of race) are maintained. This formulation seems right, but put in this way, is this problem just about freedom? In order to explain what exactly is wrong with AI creating or maintaining specific discriminations or to talk about how different groups are impacted differently by climate change and say why this is problematic, freedom, but also notions such as equality, justice, and perhaps identity will have to be used, at the very least in conjunction with the concept of freedom. For example, one could say that freedom is unequally distributed, that the oppression is unjust, and that oppression of a particular racial group by another group is wrong. Whatever the precise formulation, my point is that we are already crossing the boundaries of the liberty discussion here. It seems that liberty is not enough; we need to use other political principles instead or at least in addition to liberty. Moreover, these principles in turn can be interpreted in more universalist ways and in more particularist and identity ways.

Consider what happens if we construct the argument based on the notion of negative liberty. Either one could say that *all* individuals should be free from oppression (or that all should have equal freedom) or one could focus on the freedom of (members of) a particular racial group, which, it could be argued, should be given priority political consideration given their historical and present lack of freedom. Perhaps one could say that the freedom of one group goes at the expense of the freedom of another group. But if one tries to develop this argument and justify why exactly this is a problem, soon other terms will slip in. For example, if a particular (use of) AI discriminates against black people in the US, as Benjamin argues, then this is problematic because it is unjust and linked to other historical and present injustices, based on identity blindness on the part of 'white' universalists. This argument is made in terms of justice and identity; it is not just about liberty. Insofar as the argument is about both freedom and identity, one could call this 'identity liberalism'. But, as said, identity used to be seen as private in classical liberalism (which, as identity liberals argue, made their specific interests and problems invisible). By making the private public, identity politics crosses a line drawn and maintained by classical accounts of freedom and politics since Aristotle. For classical liberals, this line crossing is a problem; for identity liberals, this is exactly what we need to do. Similarly, for classical liberals, this argument is no longer about liberty – whatever else it is about (e.g. identity, justice) – whereas identity liberals would argue that universal freedom does not mean much if we do not consider the freedom of people belonging to specific groups (defined by identity).

A second attempt to frame the problem could start from the interpretation of liberty in terms of capabilities, where we meet similar tensions. Classical liberals interpret capabilities in a universalist way: capabilities are seen as belonging to individuals as human beings capable of human flourishing, and the goal of AI or other measures to deal with climate change is to ensure maintaining or increasing general human capabilities. And while Sen and Nussbaum have intervened in debates about justice and the justice concerning specific groups such as women, capabilities are formulated at an individual level and seem tied to the universal categories of human being and human flourishing. The starting point is individual and universal justice or freedom, but not a specific group identity. Yet once we talk about the capabilities of people belonging to these groups (e.g. defined in terms of race or gender) and compare by making claims about justice or equality, as Nussbaum does, we are again crossing a line from freedom to other values and from classic liberalism to identity liberalism and a politics based on identity. Whether or not this way of thinking about politics is a problem or a good thing in general, it is clear that classic thinking about freedom gets stretched and messed with once we deal with real global world problems concerning technology and climate and try to formulate the problem from a liberal-philosophical point of view.

Moreover, the distinction between the private and the public gets blurred when we link liberty to the good, to ethics. It is unavoidable that we move from the idea of a 'thin' classical liberal state to one about a state that is more concerned with 'thick' good of people and for society. And in ethics, too, we meet a similar tension between universalism and its critics. In the universalist view, the 'thick' ethical dimension of freedom is interpreted in the sense of needs and capabilities, not identity; the alternative view starts with (group) identity. Moreover, we do not always know what all what promoting and realizing this good and capabilities means in particular cases and in advance. But one could argue that we know *some* good, for example, that we know needs of people (universal needs) and we know universal capabilities (e.g. as defined by Nussbaum), or that we know – or should make an effort to know – the needs and capabilities of specific groups. One could also argue that both kinds of knowledge can guide our Deweyan experiments and discussions.

Thus, with regard to the good we can go in two directions. One is the classic liberal view: it may well be that there is also good that is tied to identity, but if we stay within the boundaries of classical liberalism, then whatever other good there may be, the good that matters *politically* and should be *publicly* discussed should not be tied to identity but be universal. Another, identity liberalist version is focused on group identity: here concern with the good of a particular group – defined by identity – is the starting point. Yet, for it to remain a liberal and not collectivist view, it would have to claim that the ultimate end is the freedom and good of the individual, rather than that of the group. What matters is not so much the identity and good of the group as such, but the group identity and the good of the individual.

But, in spite of these differences, there is something in common: by focusing on the good, both directions in political ethics depart from older interpretations of liberalism that privatized the good. Moreover, both directions in political ethics are relational. The universalist versions put the emphasis on our relations with others as human beings, whereas the identity version puts the emphasis on our relations to groups and relations between groups. Furthermore, one could also move beyond a strictly individualist and libertarian version of liberalism by realizing how individual good is always related to the good of others: from a relational point of view, it could be argued that good should not be tied to individuals alone; their freedom means nothing if it is not linked to the freedom of others and, in the end, to the collective (and its natural environment). According to this arguably more radically relational view, there is an intrinsic connection between individual and common good.

One could object now that this focus on the collective good and on political ethics is not really liberal. This is true if what we mean by liberalism is Hobbesian liberalism and individualist libertarianism, including libertarian paternalism. According to these views, the collective can only appear as a powerful or manipulative monster and ethics is privatized. But on the basis of the previous discussion, one could argue that we need a modified, richer form of liberalism that understands that

1 freedom is only possible under certain conditions, including the survival and flourishing of the collective (and its natural environment); for this purpose, some restrictions on negative liberty are necessary and justified;
2 negative freedom is not enough; liberty is also about what you can do with that freedom; this can be formulated in terms of needs or capabilities, for example;
3 complete freedom, that is, individual freedom that includes positive liberty, is intrinsically connected with the good life and the good society, that is, the good of the collective (and identity liberals would add: the good of particular, disadvantaged groups);
4 freedom is and should be linked to the good: liberalism cannot suffice with an extremely thin conception of good (limited to negative freedom as the only good); there is also positive liberty, which is linked to good; note that even paternalist nudging presupposes an ideal chooser who has a 'thick' idea of the good life and the good society; therefore the question is not *if*, in order to realize this richer form of liberalism, we need a collective idea about the good life and the good society, but rather *which one* we need;
5 this collective idea of the good life is not entirely pre-given or determined by a tradition, but is and should remain fundamentally open; we have to find out what the good life is and means through

experience, experiment, imagination, and discussion; and that luckily not from scratch, since there is already collective wisdom in the plurality of traditions and we can tap into that; but those traditions themselves are not enough; creative interpretation and social experiment are needed;

6 following Dewey, it is important that this inquiry is done in a participative way; this does not necessarily mean the exclusion of expertise and leadership or the neglect of tradition, but rather a dynamic dialogue with expertise and tradition, based on education of citizens;

7 following ancient thinking about politics, liberal rulers need to be virtuous human beings, preferably the most virtuous of all; the idea that we can abandon this requirement in a liberal democracy has led to disaster;

8 this implies that self-rule also requires virtue; this means that citizens need the education that prepares them for self-rule properly understood: an education that prepares for participative-experimental democracy and a moral education.

Whether or not this combination of liberalism with the ideal of participative democracy and with ancient virtue and wisdom thinking is still 'liberal' remains controversial. Skepticism concerning this is healthy for all those who care about liberty and about political-philosophical liberalism. But I have offered arguments for why and how this approach can still be construed as being about liberty. Moreover, this version of liberalism is meant to be non-authoritarian and democratic, non-paternalist, non-collectivist in an identity sense, and ethical, in the following senses:

1 *Non-authoritarian and democratic.* A democratic version of positive liberty is possible and desirable. Today and in the light of climate change, the biggest threat to freedom comes not from the kind of Rousseau-inspired thinking Berlin criticized but from (i) a form of perverted radical libertarianism, which operates by means of right-wing populism (a form of right identity politics) and exploits the collective for individual gain for a limited number of individuals and is even prepared to install a dictatorship to protect the interest of a plutocracy and racist groups and individuals; and potentially from (ii) mistrusting Hobbesians and calculative utilitarians, who may propose social reform for the sake of the climate by means of coercion – that is, paradoxically the threat comes from within the liberal tradition itself, which has been corrupted in various ways. Ironically, if liberals are afraid of a kind of eco-utilitarian authoritarianism, a green Leviathan monster, then this monster arises out of (the Hobbesian part of) their own tradition, which mistrusts people. Today it seems that we are faced with the choice between these two kinds of authoritarianism. But fortunately, there is an alternative.

Rousseau's vision may well be problematic, but it helped us to further discuss the relation between individual liberty and collective interests/common good, and helped us to explore an alternative to versions of liberalism that limit liberty to the conception of negative freedom, an alternative which is not necessarily authoritarian but – for example, with the help of Dewey – can be given a democratic twist.

2 *Non-paternalist.* Since it educates and trusts people to use their own capacity for learning, imagining, thinking, and communication to (together) find a way to promote the interests of the common good and the planet. It does not deny that people often do not act in their own best interest and that many people on this planet act irrationally. But instead of taking this for granted and coming up with a symptomatic treatment (a Hobbesian-authoritarian or laissez-faire 'solution') without looking at the roots of the problem, it tries to fix the problem by means of education and by building participative-democratic institutions that implement self-rule and support intelligent and inclusive discussion, which can then lead to people deciding to constrain themselves, if necessary at all, to create conditions that enable survival and flourishing.

3 *Non-collectivist.* It is non-collectivist if collectivism means that the collective is defined by identity and that only such a collective counts politically as opposed to the individuals that are part of it. It may still be sensitive to group identities *within* the collective (if one takes the identity liberalism direction), but in any case it is directed at the individual and common good at the level of the collective. Moreover, in contrast to communism, it values the collective without being *collectivist* if that means that the collective always takes priority over individuals, because for liberals the collective and its democratic institutions have only instrumental value: in the end it is only there for the good of individual human beings – and non-human others that also matter (see later). And, in contrast to fascism, the collective formed as a body politic is not seen as something natural but as an artifact, an *experimental technology* (but not a Hobbesian automaton) created by humans. A technology with a history, a present, and, perhaps, a future. It may or may not become a mortal automaton, something we created and which then tends to function and operate on its own. Hobbes offers an interesting metaphor here for the collective once it is shaped into a body politic. But, it is not natural and it is mortal, and it does not have the function of a Hobbesian Leviathan. The point is not so much answering fear but enabling collective survival *and* flourishing, enabling the good for all (individual) people and the common good: the new automaton and the new social contract proposed here answer to the needs of individuals, of specific groups, and of the collective, which in turn supports

individuals in answering their needs and helps them to develop and enact their (positive) freedom in terms of capabilities. But, the goal is not *primarily* the good and freedom of a specific group or the collective (*both* of which are not defined in natural terms here). Moreover, in contrast to fascism, one individual or group should never be allowed to oppress all others: it should never take away the freedom of all (other) individuals (argument based on classic version of liberalism) or take away the freedom of other groups and enjoy their freedom at the expense of that of other groups (argument based on identity liberalism). (Yet it remains questionable how liberal identity liberalism really is; it seems illiberal in so far as it shares its identity-based group thinking with fascism.)

4 *Ethical.* Questions concerning the good life and the good society are no longer privatized, as classical liberals tried to do and as contemporary libertarians try to do. Such questions are understood as linked to the political, understood by liberalism in terms of freedom – but a rich conception of freedom. My (individual) freedom, understood positively, means nothing without (individual) good, without what I can do and should do with my freedom. And collective good (or common good) is necessary to ensure 'individual' good, which was always relational and dependent good in the first place. (And perhaps identity liberals would add: our individual freedom and good depends also on that of the group to which we belong *and* – I would add, stressing the deep relational dimension of ethics and politics – the freedom and the good of the groups we *don't* belong to.) Moreover, it is justified in any democracy understood as self-rule, to demand of the rulers – that is, in principle all citizens – to be ethical, e.g. virtuous. This requires moral education.

Making this link between political liberty and ethics and the related breaking down of the private/public barrier has its price for liberals who were addicted to the libertarian version of liberalism – which was anyway utopian or only realized for a specific privileged group. According to the 'stretched' version of liberalism proposed here, freedom is no longer 'pure' and individualist (understood in a non-relational way), but instead mixed with other values and enriched with ethics and political concern for the collective, perhaps it even gives political priority to concern for specific groups in that collective. This is a considerable stretch and, seen from a classical liberal point of view, there is a 'cost': the loss of a strictly Hobbesian, libertarian view of freedom. But it is worth paying that price and doing so is now more important than ever. In the age of climate change, the question of how to live and how to ensure the survival and flourishing of the collective becomes political (or rather: it has always been political but we forgot). It is no longer a private matter and it has never been, since it becomes clear – with the help of AI and data science, next to other technologies and sciences – that my individual life has consequences for others (other humans and non-humans) and for the ecosystems of the planet. This brings us to a crossroads: do we (1) install a green utilitarian dictatorship and use AI and data to monitor and control and manipulate people, (2) do we leave things as they are and watch both the common good and the planet's ecosystems deteriorate under the rule of libertarianism, or (3) do we find another solution – a 'third way' if you like (but not to be confused with centrism in party politics) – that respects liberty but links liberty to a political ethics and still involves the governance and education of people under certain conditions? In this chapter, I have made suggestions for this third way by understanding freedom not only in a negative but also in a positive way and by linking freedom to ancient virtue ethics and to the ideal of self-rule, inspired by both Rousseau and Dewey. This conception of liberty and democracy leads us away from questions regarding the manipulation of people and the related obsession with negative liberty, and focuses more on the question of how to help people to self-regulate as a matter of virtue and (individual and collective) good. Self-regulation then means regulation as self-mastery and self-governance by the individual trying to live the good (environmental) life, but also collective self-regulation and governance, organized in a democratic, participatory, inclusive, experimental, and open way. And given the global problems we face, some form of self-rule at the global level seems necessary – even if it is not clear yet what this would mean. We will have to experiment.

Other Political Values and Principles

Now, so far, and as a matter of being charitable to those who believe political liberty should be the main political value (the intellectual exercise in this book is to *start* from liberty), I have tried hard to first remain within the boundaries of the liberal-philosophical tradition and use its resources to discuss the question of this book, which, after all, is about *freedom* in the age of AI and climate change. The main idea of this chapter was that, while negative freedom remains an important principle, we should also consider a different kind of freedom, one that is not only about 'freedom from' but also about 'freedom to', one that is connected to self-rule, one that aims at the individual but also the common, collective good since freedom is only possible and only makes sense in this way, and one that – next to being sensitive to the needs of particular groups and individuals – starts from, and is mainly focused on, the needs, capabilities, and dignity of *human beings*, who we trust are able to define and find solutions for their common problems by using social intelligence, imagination, and learning through experience and experiment. This conception of liberty, which was already stretching the boundaries of liberalism as much as possible without crossing them, can then be used and appealed to in discussions about AI and climate change. We may demand that any use of AI and any way we deal with climate change should preserve negative liberty but also promote positive liberty in all these senses. Moreover, AI can help us to achieve a more relational understanding of human being and individual liberty, which can support this approach to *freedom*.

However, there is also another route for dealing with the challenges posed by AI and climate change, which may or may not be compatible with the one sketched, and which recognizes that *next* to freedom, *apart from* freedom, there are also *other* political principles, values, and goods, which are at least *also* important or *more* important than liberty. Freedom is certainly not the only value or political principle that matters. In the previous discussion it already became clear that, even if we start with arguments from liberty, we often we have to appeal to other political principles and values, for example justice. Moreover, there are good reasons to *start* from other principles and values. Whatever is done to people's freedom, one could argue, it is important to ensure collective survival (not individual survival alone, as in Hobbes), to protect humans from harm, to 'save the planet', to realize justice, to increase equality, and so on. This chapter opened up the possibility to use the notion of positive liberty as a justification to restrict negative freedom. But, also keeping in mind Berlin's worry about using *freedom* as a justification to restrict freedom, one could argue that it is sometimes necessary to restrict the (negative) freedom of people *on other grounds than freedom*. In fact, this is what some politicians and citizens already do, for instance, when they argue for (more) taxation in the name of equality or justice, when they demand more participatory democratic institutions, or when they argue for creating the conditions for a more inclusive society without discrimination against particular groups (defined in socioeconomic terms or in identity terms such as gender or race). If one takes these routes, the main justification for political action and policies is not freedom, or at least not *just* freedom. It is another political value, good, or principle, or a set of political values, goods, and principles that may include freedom but is not limited to it. If one takes that direction, then this implies that the very political *problem* – for example, the problem regarding the governance of AI and climate change – is defined in a different way: not as a problem of freedom (alone) but as a problem concerning another issue or principle (or a set of issues or principles). For example, the problem concerning (how to deal with) the climate crisis could be formulated not as a problem of freedom but as a problem of justice. From a liberal-philosophical perspective, this raises the question whether such a shift still falls within the boundaries of liberalism or not. And for dealing with the challenges concerning climate change and AI (and other problems), the more important question is, is this a *better* approach than focusing on liberty (alone)?

In my view, it is better to also include other political principles and values, or even start with them, for the following reasons. First, and as stated, other principles and values also matter – whatever one thinks about freedom. To restrict our political and moral compass to freedom would dismiss these other principles and values. Even liberals, who by definition think that freedom is the most important political principle, can accept that other values also matter. In other words, one can hold that freedom is the most important political principle (and hence call oneself liberal) but at the same time accept some degree of *pluralism*: next to freedom, there are also other political values; other principles matter too. Second, it may be hard to describe a specific problem in a specific context in terms of freedom alone. For example, if an AI system is biased against a particular individual or group, then why not just say that it is a problem of bias, discrimination, and justice, instead of first laboring on a freedom-based framing? Now, one response to cases where a description and evaluation in terms of another principle seems to work better is to say that freedom is only of secondary importance, at most, in this particular case. The 'cost' of this option may be that liberals then care less about the problem. But, one could argue that is another problem in itself: liberals are simply mistaken when they think that liberty is the supreme value. If this is liberalism, then liberals are part of the problem, not part of the solution. We need a political approach that is more pluralist (a direction that would be more in line with Dewey) or that focuses on other principles and values, e.g. on justice or on political problems related to group identity (as identity 'liberals' would argue). This would entail leaving liberalism, at least if liberalism means that one is committed to the claim that freedom is the supreme value. Third, if we stay within liberalism: even if a specific problem *can* be described in terms of the stretched conception of freedom developed here, it may be a lot more economical and effective to directly appeal to another principle. For example, if a particular group is systematically discriminated against by a state or by a group within the state, then instead of talking about their freedom and then understanding that freedom in terms of capabilities, linking that to ethics, etc. (in the way proposed above), it may be more efficient and effective to directly appeal to justice, equality, anti-discrimination, anti-racism, and so on. Perhaps one could still hold on to the view that these descriptions in principle can also be put in terms of freedom and the freedom is the ultimate and underlying value (and hence stay liberal), but accept a different description and evaluation for pragmatic reasons. But it is clear that in such cases there is again a thin line between stretching and breaking liberalism. How much, and what kind of pluralism with regard to other political principles can and should liberals afford?

Moreover, so far it has been assumed that the only freedom that counts and the only political good that counts is *human* freedom and *human* good. The collective has been assumed to consist of humans only. But, in the light of philosophical and public discussions about the interests, value, and standing of animals, the environment, and the planet, this seems an unnecessary and undesirable limitation. Why, for example, do we tend to exclude animals and ecosystems from the collective when we discuss freedom and politics? And given that there is so much public discussion about AI, is politics really only about humans or also about technologies? Who is and should be part of the body politic? For the purpose of thinking about climate change and AI as a political problem, we should at least *ask* and *discuss* who or what should be considered as part of the collective and who or what should matter politically (and if so, in what way).

In the next chapter I will say more about other political principles such as justice and equality, which are much needed, at least *next* to freedom. In Chapter 6 I will discuss the question concerning the borders of the political collective.

References

Arendt, Hannah. 1958. *The Human Condition*. Chicago, IL: University of Chicago Press.

Benjamin, Ruha. 2019. *Race after Technology*. Cambridge: Polity Press.

Berlin, Isaiah. (1958) 1997. "Two Concepts of Liberty." In *The Proper Study of Mankind*, 191–242. London: Chatto & Windus.

Caspary, William R. 2000. *Dewey on Democracy*. Ithaca, NY: Cornell University Press.

Coeckelbergh, Mark. 2004. *The Metaphysics of Autonomy*. New York: Palgrave Macmillan.

Dewey, John. 1991. *Liberalism and Social Action*. Carbondale: Southern Illinois University Press.

Fesmire, Steven. 2003. *John Dewey and Moral Imagination: Pragmatism in Ethics*. Bloomington: Indiana University Press.

Festenstein, Matthew. 2018. "Dewey's Political Philosophy." *Stanford Encyclopedia of Philosophy*. Accessed 21 March 2020. https://plato.stanford.edu/entries/dewey-political/

Frankfurt, Harry. 1971. "Freedom of the Will and the Concept of a Person." *The Journal of Philosophy* 68 (1): 5–20.

MacCallum, Gerald C. Jr. 1967. "Negative and Positive Freedom." *Philosophical Review* 76: 312–334.

Nussbaum, Martha C. 2000. *Women and Human Development: The Capabilities Approach*. Cambridge: Cambridge University Press.

Nussbaum, Martha C. 2006. *Frontiers of Justice: Disability, Nationality, Species Membership*. Cambridge, MA: Harvard University Press.

Nussbaum, Martha C., and Amartya Sen, eds. 1993 *The Quality of Life*. Oxford: Clarendon Press.
Pappas, Gregory. 2008. *John Dewey's Ethics: Democracy as Experience*. Bloomington: Indiana University Press.
van de Poel, Ibo. 2016. "An Ethical Framework for Evaluating Experimental Technology." *Science and Engineering Ethics* 22: 667–686.

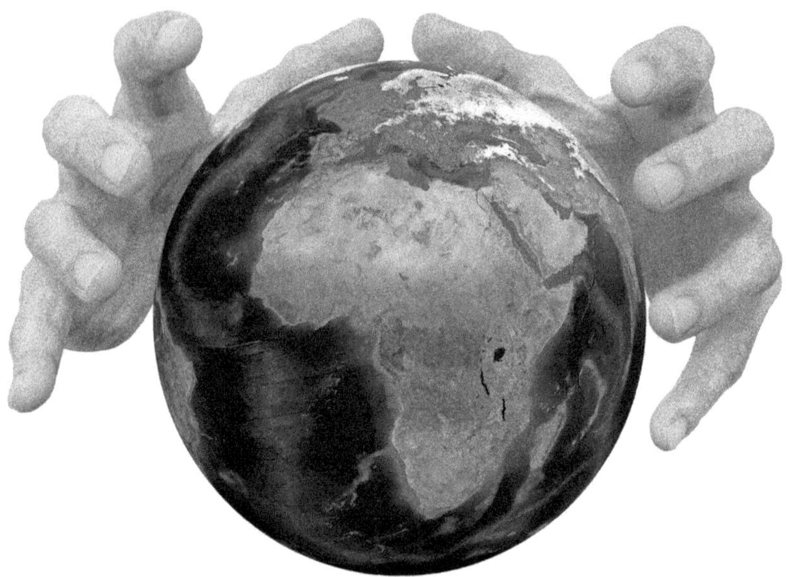

Figure 5 Invisible hands and the earth.

5 Invisible Hands in the Anthropocene

Collective Agency, Climate Justice, and the Rebellion of the Climate Proletariat

Introduction: Anthropocene, Climate Change, and AI as Political Problems

Discussions about technology and the environment sometimes include use of the term 'Anthropocene'. The idea is that we are in a new geological epoch in which humans have a dominant presence and in which human agency plays a central role. The term has been popularized by Paul Crutzen, a scientist who argued that in the Anthropocene human agency increases to such an extent that humankind becomes a geological force (Crutzen 2006). Seen in this light, the role humans play and have played in climate change is just part of a larger development that has been going on for a while, at least since the first industrial revolution: the acceleration of human power over nature. New technologies, the use of fossil fuels, and a growing world population have resulted in a planet that seems entirely under human control. This would imply that not only the present but also the future of the planet is now in our hands. As we alter landscapes, deplete natural resources, cut down rainforests, and destroy coral reefs, we create a new environment. By creating new ecosystems and by using biotechnologies, we can even create new lifeforms. We create a new nature as we transform the planet.

Moreover, just as climate change can be seen as part of the Anthropocene since it is significantly influenced by humans, AI can be linked to the new technologies and powers we have today. AI is then a tool for destruction of ancient natural environments and at the same time a tool for the management of the Anthropocenic planet. It can be used to aid the extraction of fossil fuels and thus contribute to global warming, for instance, but as the thought experiment earlier in this book suggested, it can also be used to manage the planet and people at a global scale with the aim of mitigating climate change problems. Given that human intelligence has not led to much good for people and the planet, so it could be argued, AI provides a new kind of intelligence – probably *not* a human-like one – which can help us to analyze and deal with our problems in the Anthropocene. Smart analysis of (big) data and automated actions and decision-making can help humanity to exercise its role as master of the planet, or perhaps AI could even take over from us and become a new, better master.

All this raises ethical and political questions. Is it right that we as humans take this role? Should we give the mastery and control to AI? Who is to blame for the bad effects humans and technologies had, have, and will have on the earth's ecosystems? Who is the blame for the Anthropocene? Or, framed more positively and forward-looking: given the situation and epoch we are in now (the Anthropocene), should we not take responsibility for the new world we created? Should we create a better world, and, if so, how? Should we change the way we live? In what ways? And what is the (new) relation between nature and technology, and how can we (re)shape it in a good way? For example, Paul Crutzen and Christian Schwägerl have argued that since in the Anthropocene 'nature is us', we should no longer behave as 'barbarians' who 'ransack their own home' but take responsibility as stewards of the earth and collaboratively build and invest in a new culture and infrastructure that no longer depletes the earth's wealth but instead grows it (Crutzen and Schwägerl 2011).

As we will see, the term 'Anthropocene' is not uncontroversial and this way of framing the question regarding climate change and AI can and should be criticized. But let us for now use it as a tool to discuss the future of AI and climate change as ethical and *political* problems. If there is indeed such a condition at all, then what are the implications for responsibility, agency, and power with regard to AI and climate change? For example, one could ask: if in the Anthropocene humanity has become the master of the earth, is there nothing we can do to change this and 'save the earth'? Is it already too late? What power do we have as individuals to change things given the geological force humanity as a whole already has? Is individual action enough? What does collective agency mean at a global level? And will efforts to regain mastery in order to do good for the earth and mitigate climate change – for example, by using AI – not lead to *more* Anthropocene rather than less, as I will argue?

For a good ethical and political discussion, we need to first better understand the nature of agency in the Anthropocene. For this purpose, I propose to borrow, and critically discuss, a concept and metaphor from political philosophy and economics: the 'invisible hand'. I will argue that agency in the Anthropocene can be described and discussed as an invisible hand or rather as invisible hands (in the plural), which are hidden by those who have an interest to do so, but become more visible today. I will argue that once we – helped by AI and data science – become aware of these hands, this politically significant event will most likely have political consequences, since, next to creating challenges to freedom, it raises problems concerning equality and justice (as fairness). Next to freedom, these principles are at least *also* important, in general and also with regard to dealing with global risks and vulnerabilities related to climate change.

Collective Agency and Action in the Anthropocene: A New (In)visible Hand?

As used by Adam Smith, the term 'invisible hand' refers to the unintended societal benefits of self-interested actions by individuals. Smith writes in *The Theory of Moral Sentiments* (2009) that the rich (in the 18th century: feudal lords) 'are led by an invisible hand' in the sense that, without intending and knowing it, they divide up their produce and therefore 'advance the interest of the society' (79). The more general idea that self-interest could produce the greatest good for all was then later used as a justification for laissez-faire economics and politics. The invisible hand was defined as the invisible force of the free market that leads to an equilibrium without regulation by the state. Today many economists question this idea and think it is a myth, and even Smith himself already suggested that 'natural liberty' doesn't always work. Nevertheless, the idea, based on the quasi-gothic metaphor of the 'invisible hand', remains influential in contemporary neoliberalism.

However, there is no good reason to leave the metaphor in the hands of neoliberal economists; one could liberate it from that context and use it to think about agency in the Anthropocene and agency with regard to climate change. First, applied to the Anthropocene, the concept of an invisible hand could mean that there is a kind of collective agency of humanity at a planetary level, which has consequences for the planet. Bad consequences: in contrast's to Smith's hand, it is not doing good. Instead of an invisible hand that guarantees good outcomes, the invisible hand of humanity at large leads to unintended *bad* effects on society and the planet, producing the *worst* for all. It is not in the interest of humanity. One could say that it is a negative or 'reverse' invisible hand. Second, this kind of planetary collective agency seems invisible. Invisible since we cannot easily track down who is causally and morally responsible for the Anthropocenic condition as a whole and for particular bad effects, including bad effects on the climate. Thus, in the Anthropocene we find ourselves being directed by a new, bad invisible hand, which is busy destroying the very body and environment it belongs to. We could imagine the invisible Anthropocenic master of the planet as a kind of demon or evil god.

The quasi-theological twist of the language here is not coincidental; it is connected to the very history of the metaphor and the meanings related to it. The term 'invisible hand' was used in the 17th century before Smith's usage; it referred to the hand of a devil instead of an angel. Smith's readers may also have connected it with divine providence, as Harrison (2011) argues. It seems that the planet is controlled and guided by an evil god. But now this god is humanity itself, which assumed divine powers and is playing god, having created the world as it is and seemingly having omnipotence. Yet nowhere can we see or find this 'humanity', this super-agent. The invisible hand is the hand of humanity that acts as a kind of ghostly, invisible god. It is not an automaton, like the Leviathan, that other, all-powerful agent. The automaton is visible. The monster and its organs are transparent. The invisible hand, by contrast, is not material and more spiritual or ghostly: it is the hand of a malicious god or evil spirit. The sovereign of the earth is not a visible monster, like Leviathan, but an invisible sovereign called 'humanity' or 'the Anthropos'.

The hand is also invisible because (at least so it seems) no one decided to go in this direction. There is no conspiracy. Without anyone making a decision, we somehow 'ended up' in the Anthropocene, including climate change and indeed the climate crisis. As Di Paola (2018, 126) puts it: 'we are remaking the planet and undermining the conditions of our own existence, though no individual or collective decision was ever made to do so.' There is thus a kind of collective agency that acts without a collective decision. But while invisible, the hand is very powerful, like a god. And AI seems to make it even more powerful, since the planetary agent, the master of the planet, seems to also become all-knowing. Via the collection of big data and their analysis by means of AI, humanity is now also assuming another divine characteristic: it is omniscient (all-seeing) next to omnipotent (all-powerful). Everything and everyone become transparent for it. The invisible hand can not only do everything, but it also sees everything. The invisible hand has been enhanced with cyborg-like sensors, giving it enhanced surveillance capacities. While not visible itself, it can do everything and see everything. Humanity or the Anthropos has become the invisible, malicious Guardian that rules the inhabitants of the earth, which has been turned into a perfect panopticon.

However, if this is at all a suitable metaphor to describe collective agency in the Anthropocene, we should question *how invisible the hand really is*. Thanks to increased information, research, and indeed thanks also to AI, we now see much more and it turns out there is not one ghostly hand. There are many hands, and they become increasingly visible. In other words, the invisible hand becomes plural. The seemingly monotheistic, collective, invisible meta-agent turns out to consist of many individual agents, some of which have more power than others. There are bigger hands and smaller hands. Even if the result (the climate crisis) may be unintended, one can point to specific actors whose actions have good or bad effects on planet and climate: some countries, some companies, some individuals, etc. This revealing amounts to a shift from a monotheist to a polytheist understanding of the invisible hand, and ultimately to its secularization and disenchantment. Once we abandon the gothic and monotheistic theological part of the metaphor (and thereby its conservative implications) and make visible the causal connections and social situation, we can offer a critical analysis of the Anthropocene's responsibility problem that has moral implications. Once we identify the concrete, multiple hands, companies can then be blamed and held responsible. Politicians can be blamed and held responsible. And with the help of AI and new technologies of surveillance, even individual citizens can be blamed and held responsible since their ecological and climate footprint becomes increasingly visible. AI contributes to the building of a world of total transparency. This creates the risk of AI taking over (the 'necessary evil' of a Leviathan automaton) or the danger that AI helps the evil invisible hand of humanity, but also enables us to hold people responsible for their actions and impact on climate change. Yet it also leads to the emergence of powerful agents such as private companies and governments who use this knowledge to manipulate, exploit, and govern – often covering up their own contribution. It is convenient if consumers and citizens can be blamed; if it can be shown that their hands are full of 'climate blood', perhaps nobody will see the powerful hands that operate at other levels and in other contexts. Powerful actors may prefer that we speak of the invisible hand of 'humanity' or the 'Anthropos', since it renders their own actions invisible. It is also convenient for them to say, in a perversion of Dostoevsky's moral existentialism, that 'everyone' is responsible. This then again constructs the fantasy of a collective agent understood as an invisible hand, which diverts attention from the fact that some agents are more powerful and have a larger hand in the bad (and also potentially good) effects. Again, the *invisible* hand concept seems to support neoliberal laissez-faire: if it is humanity at large that is the bad agent, then this keeps the problem sufficiently abstract and we can let companies, nation states, and individuals do whatever they want. Then it is not their problem. The politics of the Anthropocene, understood by means of the metaphor of the invisible hand, thus involves a politics of covering up and revealing – not unlike a divine drama, a striptease, a magic show, or a gothic story. But these ways of thinking and narratives are not just fictional; they have real moral significance and implications.

Indeed, once we lift the veil of the invisible hand implied in the 'Anthropocene' narrative and reveal the field of agency and power that was covered by it, we can see that there are good reasons why some actors are more blameworthy than others. Winner's (2017) criticism of the 'Anthropocene' concept supports this point: he argues that the term 'Anthropocene' hides that many human beings 'have lived modestly with minimal impact on the local or global environs or the Earth's climate systems' (285); some people contribute far more than others (e.g. capitalists) and some generations more than others (287). We should question our 'actual social and economic institutions and activities that are clearly the primary cause of the massive effects in the biosphere evident today' (Winner 2017, 285). In other words, we can re-describe Winner's argument as an argument for making the invisible hand visible. The term 'Anthropocene' was hiding a lot; it is time to reveal this now. It is time to do away with the invisible hand metaphor that – implicitly and without our realizing it – has powered the Anthropocene concept, and to reveal the patchwork of agency that was covered up by the 'collective agency' it suggested.

However, questioning the kind of collective agency suggested by the metaphor of the invisible hand does not necessarily mean that we have to revert to a political or methodological individualism. Critical discussion of the invisible hand metaphor also helps us to problematize the individualism that is often operative in discussions about climate change (and indeed about technology and AI). 'Individualism' can mean various things here. First, it can mean that for the analysis of moral responsibility the focus is on the ethics and politics of individual citizens rather than other actors. As I already suggested, individuals are blamed, whereas companies and governments protect themselves from blame. Given that the latter have a lot more power and thus have a larger causal impact on climate change, this seems wrong. More powerful actors should carry more responsibility for what goes wrong – and of course also for doing the right thing. This takes us to the forward-looking question concerning responsibility. Second, 'individual' can be opposed to 'collective' in a way that suggests that only individual action is needed to deal with global problems such as climate change, whether by individual citizens or by other individual actors such as companies or nation states, and that (more substantial forms of) cooperation and collective action is not desirable. There is often a neglect of collective solutions, and this is partly intended by some actors. Those who believe in individualist and neoliberalist solutions hope that individual actions lead to a new, benign invisible hand. Collective action is rejected or not even mentioned, and cooperation is minimized, also and especially on a global level. But this will not be sufficient; we need collective action and collective solutions given the global and challenging nature of the problem. Crutzen and Schwägerl talk about the need for a 'collaborative mission that dwarfs the Apollo program'. And going further than cooperation in the sense of inter-individual and international cooperation: in order to deal with global problems such as governance of climate change and AI, one could argue that we need to set up a political institution that can exercise collective agency at the global level. But, in contrast to the invisible hand of humanity in the Anthropocene, this supranational agent would have to be a *visible* and a *benign* 'hand'. Note, however, that one may agree that collective and collaborative action is necessary to deal with the problem, but disagree about the precise form that it would have to take. For example, can it be done in a way that preserves freedom and democracy? (I will continue this discussion later in this chapter.) And what is the role of technology?

It is clear that collective and collaborative action may include technological solutions. But which ones? Crutzen and Schwägerl propose that we develop geoengineering capabilities. One could also add AI to the climate mitigation toolkit. However, from the point of view of trying to deal with the 'Anthropocene' problem, such solutions may well contribute to the problem rather than solving it, since they increase humanity's grip on the earth. If a technology increases the agency of humanity, then it creates more Anthropocene rather than less. In response to this problem, one could question the stress on action altogether, and recommend to try to loosen our grip. Heidegger seems to suggest such a direction with his term '*Gelassenheit*' (Heidegger 1966), as I have argued elsewhere (Coeckelbergh 2015, 2017). The term refers to an attitude of letting go or releasement. A kind of releasement from technology, which Heidegger proposes, could be accompanied by a releasement of the earth. This can be seen as a critical response to technological solutionism and an exclusive focus on action. And in contrast to what critics of Heidegger may fear, it is not necessarily a conservative response. Influenced by Bateson, Marcuse, and Taoism, Van Den Eede has recently argued that the solution-oriented 'act now!', such as neoliberal calls for concrete action, should be greeted with skepticism. He notes that it is often a way to maintain the existing situation and is often conservative (Van Den Eede 2019, 117). Loosening our grip on the earth, letting go of the earth, might be a good and progressive response to the Anthropocene.

The Hobbesian Temptation (Again)

However, this reasoning still takes the Anthropocene problem definition and its implicit use of the invisible hand metaphor for granted. Moreover, does this emphasis on collective action or releasement mean that individual action in a good direction is not desirable? If we want to use the metaphor at all, would a *benign* and *invisible* hand be possible and desirable, next to the benign visible hand of collective action and cooperation I just proposed? This depends on the conditions. In a Hobbesian world that supports competition, self-interested actions will not lead to good for people and the planet. From a Hobbesian point of view, one may remain pessimistic that this condition will ever change. But, following (my interpretation of) Rousseau again, we could also trust people to do good for the planet and people if and once the right social and institutional conditions are created, conditions that minimize competition and stimulate work for the common good. Then we could believe in the emergence of a benign invisible hand that, through many individual (in)actions, does good for the planet (or leaves the planet alone). *Under such conditions*, an invisible hand may well arise and could then work together with the visible hand of collective action and cooperation proposed in order to deal with climate change.

However, one could argue that, since presently these conditions are not fulfilled, we cannot trust that an invisible hand will lead to a collective benign agency for planet and people (or will decrease humanity's agency and grip on the planet). Therefore, given the persistence of a Hobbesian situation – at the global level and at least and especially with regard to environmental and climate politics – one could revert back to Hobbesian reasoning and argue that we need *first* an authority that steers the collective and takes collective action. This *visible* hand of collective action and intense cooperation seems necessary for survival. Then later (and preferably in the meantime) the social order should be changed in such a way that the right social conditions are created so that an environmentally benign invisible hand can emerge and do its work. First we need to reach the goal of survival (and for that purpose less negative freedom may be necessary), then and in the meantime conditions for positive freedom can be created (e.g. through education), leading to people taking care of the planet and people through self-rule. Thus, one could argue that, when it comes to dealing with climate change, and before we can realize Rousseau's political ideal, we need Hobbes's solution to deal with a Hobbesian situation.

Now, one could argue that some countries and communities are more removed from a Hobbesian situation than others. Some already to some extent created conditions (or did not completely destroy historical conditions) that led to some trust and interest in the common good. In such countries, then, an invisible benign hand may well emerge, or has already emerged. But whatever the situation in particular countries, at the global level we still have a more or less Hobbesian situation. This seems to imply, unfortunately, that in the short run we need a Hobbesian solution – while we keep working on changing the conditions and creating a non-Hobbesian world. If we refuse this, conditions for survival are threatened, which means that creating conditions for the good life and the good society does not even get off the ground.

However, this Hobbesian argument is misleading insofar as it falsely suggests that we find ourselves in a global state of *nature*. The competitive, non-cooperative, and nearly radical libertarian situation with regard to environmental and climate politics we find at the global level (and often elsewhere too) is not a 'state of nature' in the sense that is the 'natural' order, 'before' the social order. It is artificially created. We created a world of competition in which individuals, companies, and states can more or less do what they want when it comes to care about and for the environment and dealing with climate change. And if it is artificial, we can also change it. We can create a different, non-Hobbesian world. In other words, one can use Rousseau again to argue against a Hobbesian problem definition.

Furthermore, if collective action takes the form of a Hobbesian solution, it leads to a form of authoritarianism. According to Hobbes, it even entails absolute power. The argument for installing authoritarian rule is that otherwise individuals (Hobbes) or states and (I added) multinational corporations will still do what they want. Now one could question this assumption. One could also question Hobbes's view that absolute power is needed to solve the problem. Perhaps the question how much power is needed to ensure survival is in the end an empirical question. But it is clear that, within this Hobbesian thinking, an authority that is powerful *enough* needs to be established. In my argument: at the global level. And if this takes the form of authoritarianism, this remains of course problematic from a democratic point of view – which I have employed against Hobbesianism in the previous chapter. Moreover, saying that 'first' we need authoritarianism to 'solve' things and then 'afterwards' we can have democracy is also very problematic since in practice this argument is often used by authoritarian regimes to justify the continuation of their rule. Usually the 'afterwards' doesn't come.

So what can be done to avoid going in a Hobbesian direction? I already proposed to create conditions for a non-Hobbesian world. However, because of the goal and value of survival (of individuals and of the collective), it was tempting to argue for a Hobbesian solution in the short run. Survival is of course an important value. It is not only promoted by neoliberal or fascist Hobbesians, who propose that we compete or fight for individual survival (or group survival); survival is also the supreme value of 'left' and 'ecological' political movements such as Extinction Rebellion: what counts for this movement, primarily, is the survival of humanity and the survival of the species. But what happens if we question the premise that survival is the only and main political goal and value? What happens if we change this assumption? Perhaps this is our only chance to move beyond a Hobbesian approach and to avoid eco-authoritarianism at a global scale, avoid a green Leviathan.

Beyond a Hobbesian Approach: From Survival and Freedom to Climate Justice and Climate Equality

Liberty versus Other Principles: Justice and Equality

Against the Hobbesian problem definition and questioning the focus on liberty in the forms of liberalism discussed in the previous chapters, one could argue that what matters with regard to the climate crisis is not so much and not only survival or liberty but also *other political values, principles, and goods.*

For example, faced with the new vulnerabilities, risks, and indeed already ongoing harmful consequences of climate change, one could argue for a redistribution of risk and negative consequences, and justify this by relying on conceptions of *justice*, in particular justice as fairness and distributive justice. If we shift our concern to climate justice, understood as climate fairness, then we can and should talk about whether the distribution of risk and harm related to climate change within the collective is fair and *then* (if necessary at all) discuss what this means for liberty. The main worry here is not that humanity as a whole will perish (together with many other species) but that some humans – some individuals, some groups, people from some geographical/climate areas – will survive whereas others will not (and similarly that some species will survive whereas others will not). And going beyond the survival question: if one is concerned about climate justice, one may worry that some humans and non-humans will benefit whereas others will carry the burdens and costs of climate change – risks and negative, harmful consequences. One could argue that if and in so far there is an unequal distribution of vulnerabilities and risk with regard to climate change, then next to minimizing existing harm and working toward other survival-oriented goals, such as supporting the survival of species and increasing biodiversity in general, it is from a justice point of view important to evaluate the justice and fairness of that distribution and, if necessary, redistribute that kind of risk and vulnerability.

Of course one could and should consider the consequences for *freedom* of such a redistribution. Will people who are advantaged in terms of climate change vulnerabilities be *forced* to give resources to others or to the (global?) state in order for those others to be less vulnerable to climate change? What is the cost of more justice in terms of freedom? And one could also ask further questions in terms of freedom: how free are people who currently have to pay the costs of climate change (more than others)? Perhaps they have negative liberty, but not positive liberty in terms of meeting their needs and enhancing their possibilities and capabilities. And do people have freedom as self-rule to decide about all this, or will an authoritarian ruler or group of experts decide the distribution of climate vulnerabilities and risk? Even someone who thinks that justice is more important than freedom could and should ask these questions, assuming that liberty remains *a* political value. However, here the point is *that the political question regarding climate change is not, and should not be, only about liberty.*

Another political principle and concept is *equality*. Some humans (and some non-humans) are more vulnerable than others. This problem can be formulated as being primarily about equality rather than justice. If there is an *unequal* distribution of vulnerabilities, then one could ask if this *inequality* is acceptable, rather than asking: is it just and fair? Furthermore, as in the case of fairness, this raises the further question: if it is not equal, then what should be done? Should we redistribute climate risk and vulnerabilities, is this an acceptable means to reach the goal, and how? What kind of equality do we want? What does equality of climate vulnerability mean in practice? And if we implement measures to increase equality, what is the cost in terms of freedom? How much freedom do we want people to have? Or is freedom not necessarily to be seen as a threat to equality but, instead, is some form of equality required for (positive) freedom? Answers to these questions will involve views about freedom but also about what an equal society is. Again, the point is that not only survival or liberty count, politically. There are also other political values and principles.

Note that both arguments in terms of justice and equality can be linked to more universalist notions, for example, human rights or capabilities rooted in human being and human flourishing, or can be formulated in particularist terms, for example, in terms of group identity. The direction depends on one's view of *to whom* justice should be done: to individuals as human beings or to individuals and groups with a particular identity (or individuals as part of groups with a particular identity)? At the global level, that is, in international institutions, there is a tendency to focus on universalist notions of justice and equality. One may well prioritize justice for specific individuals and the regions from which they come, based on a conception of justice that prioritizes the least advantaged in climate terms or wants to make sure that they have sufficient protection against the effects of climate change (see below for principles of justice), but the philosophical underpinning is universalist. Justice needs to be done to all individuals because as human beings they are all intrinsically valuable, have human dignity, deserve justice and equality, etc. – as human beings, not because they have a particular individual or group identity. The alternative is a politics based on identity, according to which group identity matters when it comes to climate justice and equality, for example, because a present (climate) injustice to an ethnically defined group is linked to other (present) injustices done to this group and/or to historical forms of injustice in the context of colonialism.

Note also that there is an important link to technology here, including AI. Like other technologies (Coeckelbergh 2013), AI can be used to tackle climate problems, but is also likely to create new climate vulnerabilities or at least increase existing ones. For example, it may be used for the optimization of extraction of fossil fuels, destruction of forest, etc., and thus contribute to depletion of natural resources. This will have an effect on humanity at large but also on some individuals, groups, communities, and parts of the world more than others. Again, the problem is not only that humanity at large becomes more vulnerable, but also that specific groups and individuals suffer *more* harm than others and are more vulnerable, for example, when global warming makes it more difficult for some people to make a living in the way that they were used to and when they cannot protect themselves from 'natural' disasters and extreme weather partly due to global warming.

In response, one could decide to protect the most vulnerable and, more generally, foster climate justice by redistributing climate vulnerabilities. One could justify such a redistribution by appealing to the principles of justice or equality. But here we hit upon the problem of freedom again. Redistribution of risk and vulnerability (as caused by use of technologies) requires an authority. Even if the goal is no longer, or not only, Hobbesian survival, supporting other ethical values and political principles such as justice and equality also requires an authority. Perhaps not an authority with absolute power, and hopefully a democratic one, but, in any case, one that has enough power to enforce the redistribution, at the global level and at lower levels. This will necessarily lead to less negative freedom on the part of individuals. But it will ensure more justice (e.g. justice as fairness), more equality, and so on – with regard to climate change, AI, and potentially also other issues. This raises the question of how to balance freedom with other values such as justice or equality.

Another route, however, and one which can be followed in conjunction with redistribution, is to *not take the technology for granted* but change the technology and the practices it is related to in order to reach goals regarding justice, equality, or freedom. In this case, one could develop and use AI in ways that do not have the bad effects in terms of harmful consequences or increased risk and vulnerability. However, this again requires national and global governance, probably to a much greater extent than is the case now – which again raises the problem of freedom. If we really care about preventing the use of bad (uses of) technology, and, for example, regulate AI in order to prevent bad effects on climate change, what is the cost in terms of freedom and is this acceptable? And how can we stimulate and ensure the development and use of *good* technologies? Instead of only asking a Hobbesian police state style question (about prohibitions that are meant to avoid bad outcomes) one could also ask: what is the positive ethical and political goal or principle we want to reach with this technology? For example, one could govern AI in a way that stimulates the development and use of AI that contributes to climate justice.

Similarly, questions regarding fairness and justice, equality, freedom, etc., should not only be asked when it comes to risks and negative consequences. The question of the *good* life and the *good* society should also be connected to it. We might know what we do not want. But what is the positive political and ethical project? What is a good and just society? What is a good way of dealing with climate change? What is a just and fair way of doing this, for example? Again, it may turn out that to realize this requires a redistribution of some sort. If it is desirable, as I argued, to create conditions for positive liberty, capabilities, and the good life and good society, then creating such conditions may also involve some kind of redistribution, for example, a tax system that creates flows of financial resources from the rich to the poor.

Climate Justice: Principles of Distributive Justice, Intergenerational Justice, and a Discussion of Libertarian Objections

But what is a fair distribution, what is a fair distribution, for example, of risk? Political philosophy offers various ideas about *distributive justice*, which we can also apply to the question regarding climate justice and risk. Theories of distributive justice offer principles that tell us how goods, risks, or other things should be distributed. The two most influential principles of distributive justice are sufficitarianism and prioritarianism. Sometimes egalitarianism is also included, but strictly speaking this belongs to theories that prioritize the principle of equality over justice (see below).

Sufficitarianism holds that everyone should have enough of a particular good. For example, one could hold that everyone should have enough money or a sufficient level of a particular capability. It is not important that everyone is equal, in the sense of having the same, but that everyone has *enough* (Frankfurt 1987, 21). For climate justice, as I defined it in terms of risks and vulnerabilities, this principle could translate into the demand that everyone should have *sufficient* protection against the risks of climate change and *sufficient* reduction of climate vulnerabilities, regardless of relative differences in society (or at a global level). Sufficitarianism proposes a minimum level or threshold, which here would take the form of a protection against climate risk threshold. What happens above the threshold is of no concern to sufficitarians when it comes to justice. Some may have a higher level of protection. What matters is the (absolute) level of advantage. *Prioritarianism*, by contrast, is interested in relative differences and holds that when giving benefits, priority should be given to those who are worst off. In the climate justice case, priority would be given to those who are most vulnerable to the risks and effects associated with climate change (and AI). Some believe that John Rawls's so-called 'difference principle', according to which 'social and economic inequalities are to be arranged so that they are to the greatest benefit of the least advantaged members of society' (Rawls 1971, 266), is a prioritarian principle. Applied to climate risk, using a difference principle would mean that an unequal distribution of risks and vulnerabilities is fair if and only if it is for the benefit of the worst off in terms of climate vulnerability: those who are most vulnerable in terms of the consequences of climate change. Note, however, that this is still formulated in terms of justice (as fairness). Prioritizing the most vulnerable may well have the effect of decreasing equality in society (or at a global level), but equality is not the aim. *Egalitarians* will argue against this approach: they will say that climate change and its associated risks and vulnerabilities is not a question of justice but equality, which for them has not instrumental but intrinsic value. They may demand equal conditions for everyone, e.g. equal risks and vulnerabilities with regard to climate change. For them, equality is desirable per se. There is a lot of discussion about what egalitarianism means (Arneson 2013), and, for example, much depends on what exactly is equalized: equality with respect to what? Here one could propose equality in terms of climate risks and vulnerabilities.

Given the topic of this book, one may ask what realizing these principles means in terms of freedom. For example, leaving aside principled objections that can be brought against egalitarianism from the side of prioritarians or sufficitarians, realizing *absolute* equality seems to require more centralized power and perhaps even a form of authoritarianism, since in a Hobbesian situation a lot of power in the hands of an authority is necessary to ensure absolute equality, whereas reducing inequality (without leveling out) or justice as understood in sufficitarianism or prioritarianism may still require sufficient authority but leave more room for negative freedom – depending on the threshold or on how much resources are redistributed to those who need it most. In any case, one or more of these principles can be used to define what justice is and what a fair distribution is. Furthermore, whatever principle is used, there needs to be a discussion about *what* needs to be distributed or redistributed. Is it financial resources? Is it risk or vulnerability, as I suggested? But what, exactly, does it mean to redistribute risk or vulnerability? What is needed for this? And, not to forget: *who* should do something about it, who has the duty to redistribute for instance? It could be 'individuals, corporations, states, international institutions, or generations as a whole' (Page 2007, 18). I would add 'supranational institutions' to this list.

Climate justice can also mean that one considers the distribution of climate risk or other things *in time*, for example, between generations. This is so-called *intergenerational justice*. Tackling climate risks and vulnerabilities by reducing them and potentially redistributing them is then seen as a way to address an unfair distribution between generations. For example, one could claim that future generations should not carry the burden of our current climate-impacting actions because this would be unfair to these future generations.

Note that an intergenerational claim could also be made in terms of freedom, rather than justice, at least if we define freedom in a positive way. Sen (2004) has argued that we should preserve the freedoms of people today 'without compromising the ability of future generations to have similar, or more, freedoms' (1). Sen and Nussbaum define these freedoms as capabilities, which I have interpreted as a notion of positive freedom. The notions of capabilities and positive freedom can thus be connected with justice, as Martha Nussbaum has done for capabilities (interpreted in a universalist way): justice demands that all human beings be able and enabled to develop their capabilities (Nussbaum 2007).

Applied to the problem at hand and giving it an intergenerational twist, one could claim that climate justice demands that people's capabilities with regard to climate change should be preserved and enhanced without compromising similar capabilities of future generations. Finally, from an identity politics point of view, intergenerational justice can also mean that the current generations of a specific group (defined in terms of its identity, for example, defined in racial and historical terms) are asked to compensate for injustices done in the past by their ancestors to people of another group. For example, if some of the present climate injustices and risks were partly created by colonialism, then the descendants of the colonial oppressors would have a special responsibility to address these injustices and do justice to the descendants of the oppressed – perhaps even if the latter are no longer oppressed and no longer suffer from open forms of colonialism and its injustices (although often discriminations still continue).

Now all these different approaches to justice and equality assume that issues regarding climate risk and vulnerability are mainly a question of justice or equality. Laissez-faire *libertarians* à la Austrian-British economist and philosopher Friedrich Hayek would contest this. They will argue that dealing with climate risk and vulnerability is not a question of redistributive justice or equality at all, that these approaches lead to measures that infringe negative freedom, and that since liberty is the supreme value and principle, these infringements cannot be justified. They could argue that since *only* freedom is important or since freedom is *more* important than justice or equality, it is morally and political acceptable that only the well-off survive, that some people do not reach a minimum threshold of protection against climate risks, or that the already socially and environmentally vulnerable (e.g. poor people, people in the global south) carry the biggest burden of climate change. They might still regret this situation, but then appeal to the invisible hand argument and say that actions of the advantaged (still) have benefits for those who are less advantaged, thus rejecting any measures by the state taken in the name of justice or equality. Note, however, that Hayek himself was not against collective action as such, he even argued that this would not restrict individual liberty (Hayek 1960, 257–258). But it is not clear how this is possible. And presumably such collective action could then only be taken in the name of freedom.

Against this (extreme) libertarian objection, either one can endorse the view that another principle (justice or equality) is supreme or one can take a pluralist approach as suggested earlier, and argue that liberty is a value but not the supreme value, and that therefore justice and equality are also important. Moreover, when more (big) data become available and are analyzed by using AI, it may become increasingly clear that the invisible hand does not work when it comes to dealing with climate change, except in a negative way (making things worse when it comes to climate change). It may also become more visible who is doing more damage to the environment and the climate and who less, and who is suffering most from climate change and who less, who carries most risk and who is most vulnerable and who is less at risk and less vulnerable, and so on. What could be the political implications of such enhanced social transparency?

Rebellion, Revolution, and Fatalism: Scenarios for the Climate Proletariat and AI as a Challenge to Humanism

Class Struggle and Revolution: The New Climate Classes

Increased knowledge and awareness about this situation is likely to have profound effects on social orders that have so far relied on a laissez-faire libertarian narrative and invisible hand myth. It might shake those orders to their foundations because their social contract was based on invisibility: a kind of 'veil of ignorance' (Rawls 1971), not in regard to social-economic position and not meant to lead to accepting principles of justice, but a veil that was hiding the distribution of agency and benefits of contributions to climate change and thus to the creation of climate vulnerabilities. Once it becomes clear who are benefiting most from global warming and who contributed most to it, the most disadvantaged may no longer accept to be at the bottom of the climate risk pyramid and protest and rebel. The situation of total transparency created by AI – transparency about the social order in general and transparency about climate vulnerabilities in particular – may thus lead to what Marxists call a 'class struggle'.

In classical Marxism à la Marx and Engels, a class struggle happens because in a capitalist socio-economic system those who own the means of production (e.g. industrial machinery), who are a minority, control labor and the workers (the majority). This tension leads to the workers demanding social and political change. The 'proletariat' no longer accepts that they are controlled by 'the bourgeoisie'. In *The Communist Manifesto* (1992), Marx and Engels write that the history of societies has always been a history of class struggles between 'oppressor and oppressed', which sometimes ended up in revolution (3). In modern times, this history continues: the rise of the bourgeoisie has developed a class of laborers 'who must sell themselves piecemeal' like a commodity (9). By means of machinery and division of labor, the 'proletarians' become 'an appendage of the machine' (10). Exploited by the bourgeoisie, the proletariat, which increases, struggles with the bourgeoisie. There are occasional revolts and riots (12). Communists – the famous 'spectre' the authors talk about in the beginning of their manifesto (2) – then help to politicize this struggle: they make the proletariat aware that they are a class and help to overthrow 'the bourgeois supremacy' so that they can conquer political power (17). This takes away freedom from the bourgeois (20) but liberates the proletariat, that is, the majority, from oppression by the minority. This leads first again to the rule of one class (the proletariat this time, which takes over the power) but then, in the end, to the removal of the very conditions that produced the class struggle. Marx and Engels argue that the result is a class-less society: 'an association, in which the free development of each is the condition for the free development of all' (26).

One could argue that a similar process will or should happen with regard to the present situation. But now, in the light of climate change, the social classes are defined differently. The difference is not primarily one between those who own the means of production and who do not, as in Marxist theory. There might still be a class difference in this sense. Today, for example, a few corporations own many of our data and have access to the best technology (e.g. the best AI). And even if data and AI were to become increasingly accessible, this accessibility or ownership would not necessarily be translated into positive freedoms or capabilities. Moreover, following Marcuse (1964) we could also put the problem of *consumption* on the table: the problem, especially with regard to the environment, is not only domination by those who own the means of production, but also a new kind of unfree, quasi-authoritarian domination via consumption. In a consumer society, the majority is unfree because they are manipulated into having new needs in order to increase consumption. This is an additional problem, which adds to the existing class difference and struggle between capitalists and the proletariat. It also contributes to the depletion of natural resources and, in the end, to the Anthropocenic situation.

However, one could argue that there is now *a new kind of class difference*: one between those who reap the benefits from doing damage to the environment and from co-causing global warming and those who carry the burdens: the risks, costs, and harms. The line between the two classes may not (yet?) be so clear and there is likely to be some overlap with classes defined in terms of ownership, but my point is that it is a different line that is drawn here, one between two 'climate classes': the climate capitalists and the climate proletariat. The former may or may not *own* the means of production, but in any case employ and benefit from technologies that have bad climate effects, and carry the least risk when it comes to the effects of climate change. The latter are most disadvantaged and vulnerable when it comes to those effects, and have given their data to the climate capitalists by using their technologies. The people who belong to the climate proletariat carry most of the heavy climate burden: they face old and new risks and vulnerabilities created by the actions of the climate capitalists, and are perhaps already harmed by climate change. Awareness of this class difference could then lead to rebellion by the climate proletariat and potentially to the end of climate capitalism, albeit not necessarily so (here I propose to take distance from Marxism and Hegelianism, which presuppose a pre-determined course of history). In any case, the class struggle will put pressure on the climate-capitalist system and those who try to justify it by appealing to liberty. Here are some scenarios:

1 *Progressive and democratic change without revolution.* The ruling class (climate capitalists and those serving them) agrees to change the social order and is willing to subject the economy to sufficient regulation; at the local and global levels, climate change is dealt with in the best possible way and climate capitalism is abolished or transformed into a better social order in a democratic way: one that benefits (all) people and the planet.
2 *Status quo and disaster.* This scenario is unlikely to happen, since the climate capitalists and those who defend them benefit from the status quo. They have an interest (albeit a short-term interest) to continue under the present social arrangements. The result is disaster.
3 *Insufficient change and disaster.* More likely is that, under moderate pressure from climate protests, the ruling class accepts *some* governance and regulation, including at the global level, but not enough to ward off the worst consequences of climate change, and certainly not enough to protect those that are most vulnerable to these effects (the climate proletariat). Environmental and (or rather: including) human disaster is the result.
4 *Revolution and authoritarianism.* Rebellion happens and climate communists succeed in overthrowing the ruling class (the climate capitalists), which results in *climate authoritarianism*, which effectively deals with the climate crisis but at the cost of destroying (negative) freedom (green Leviathan scenario; compare with 20th-century nation state communism as an authoritarian and totalitarian system).
5 *Revolution and true democracy.* Rebellion happens and (perhaps with help of a new political movement) the rebels succeed in overthrowing the ruling class, which results in the establishment of true *democracy*, which effectively deals with the climate crisis and manages to do so in a democratic way and while respecting liberty and other important political principles. Alternatively, one could also frame this scenario as one that leads to true communism, realizing the kind of liberty envisaged by Marx and Engels. This scenario seems unlikely; efforts to realize it usually end up in scenario 4.

The apocalyptic and authoritarian scenarios seem the more likely ones, unfortunately, unless we make an effort to realize 1 (democracy through reform) or 5 (democracy through revolution). In terms of justification, these democratic projects could draw on some of the principles, arguments, and discussions about positive freedom, democracy, justice, and equality constructed in the course of this book. But as I have shown, there is no clear and straightforward recipe: all positions raise new points for discussion, for example, in terms of threats to freedom or tensions between freedom and other political principles. There also needs to be further discussion about the relation between liberal democracy and Marxism (and critical theory). In order to explore a Marxist direction, I borrowed the class struggle idea; but it is not clear how class struggle and rebellion would lead to the free society Marx and Engels envisaged. In any case, reading Marx and Engels invites us to think about the problem of climate change and AI as a political problem in a way that takes seriously the political and the problem of freedom *as a social history* (here a history of class struggle) and in a way that once again at least *stretches* liberalism. Again, the question comes up: how far can we stretch liberalism before we break it? Communism in its authoritarian form (scenario 4) breaks it. Is a democratic form possible, one that deals with climate change but at the same time remains democratic and retains freedom (or realizes freedom for all)? But what does this mean? Within the critical theory tradition, one could consider a social-democratic form of socialism. But what if there is a tension between freedom for all and mitigating climate change? What if democracy – at least as we know it – cannot do the job? A social-historical approach is helpful to think about what kind of future we want, but the goal remains vague and the problem of freedom in the light of climate change does not leave us.

To conclude, discussing the politics of climate change and AI leads us to key issues in political philosophy, including discussions about the 'invisible hand', about justice, and about the history and future of our (Western) social orders and democracies. Of course, this is a selection. Let me identify at least two areas of political theory that have been left out. First, this discussion could and should also be expanded to other, less Western social orders and approaches; it would also be interesting to look at other values and principles such as compassion, for example. Different (political) cultures also have different ways of dealing with climate change and technologies such as AI. Here I have limited my discussion to questions and resources from Western political philosophy. Second, both classical liberalism and Marxism assume that the political question regarding freedom and the question regarding power are only concerned with the relation between states and their citizens. This is problematic in at least two ways. First, and as I have suggested throughout this book, there is also the question of global governance, and this (a) does not necessarily have to take the form of a modern nation state expanded to the global level and (b) in any case goes above and beyond the level of the nation state. Unfortunately (given global crises and in spite of so much scholarly activity in political philosophy), we are still only at the *beginning* of thinking through what freedom, power, citizenship, etc., mean at a global level. Second, going *below* the level of the nation state, there are also political questions, including questions about freedom. From an identity politics point of view, for example, one could talk about the open and less open political struggles between different groups, defined in identity terms, for example, in terms of gender or race. And thinking in terms of nation states often covers up regional differences and tensions between different regions. Current nation states often could only arise as a result of (what we now define as) internal colonization and suppression of regions, and often centralist states still oppress political activity on the regional level. More work is needed to discuss what this means for dealing with climate change and AI.

But there is more going on in terms of politics below the level of the national state. The French philosopher, social theorist, and historian of ideas Michel Foucault has famously argued that politics and power also take place in institutions such as hospitals and prisons and even in the family. Power, he argued, is not only in the relation between state and citizen; it is also present and exercised throughout the social fabric, across all kinds of social relations. Power is not only a matter of macrolevel systems such as capitalism, as Marxists argue; it is omnipresent (Foucault 1998) and thus present at all levels. There are 'micro-mechanisms of power' (Foucault 1980, 101) in all kinds of social relations and institutions. Applied to the use of AI for surveillance, for instance, this means that we should consider not only how states employ the technology for surveillance (say, as they exercise power hierarchically or vertically), but also how surveillance is happening on a 'horizontal' level: people exercise surveillance toward one another and even discipline each other, for example, via social media (apps). Foucault also studied ancient practices of the self, for example, in his histories of sexuality (e.g. Foucault 1998). It would be interesting to explore what all this means for dealing with climate change. What happens to our problem of freedom when we consider the question concerning climate surveillance in the family, for example, when family members closely watch each other's 'climate behavior' when it comes to food? How does 'climate disciplining' (and punishment) happen at work, for example, when workers are disciplined by management *or by each other* not to use plastic bottles? Is disciplining and punishment going on when someone sees that a colleague is traveling a lot and says that this is bad since it contributes to climate change? And how do we and should we *self*-discipline in the light of climate change? How do we operate on our bodies and souls (Foucault 1988) when, motivated by the desire to help with tackling the climate crisis, we change our diet, use a different means of transportation, and so on? What does human freedom mean in such contexts and practices? This question opens up an interesting direction of research, which has been ignored in the previous discussions oriented toward mainstream political theory. (Elsewhere I have touched on it, although not in relation to climate change; see Gabriels and Coeckelbergh 2019.)

Is the Future in Our Hands? AI as a Threat to Humanism

Using the Marxist device of class differences (but also Foucault's thinking, intergenerational justice thinking, approaches that problematize colonialism and neocolonialism, etc.) alerts us to the time dimension: discussing the politics of climate change and AI is not only about analyzing and evaluating a given and fixed social order; it is also about movement and movements, histories and futures, changes and transformations. The body politic is mobile and dynamic. It is in motion. In the next chapter I will use the term *becoming*. But already at this point we can ask questions regarding the direction of the changes. And we can ask about human *freedom*. Do we know the changes beforehand, without being in a position to alter the course of (climate) history? Or do we still have some room for decision-making? To what extent can we still make our own (climate) history, given that AI gives us more knowledge about (patterns of) physical and social processes? What if climate change can no longer be stopped? What if some social processes – e.g. toward totalitarianism – are not like physical forces but nevertheless have their own dynamics and are at least *difficult* to stop? In what sense *are* there invisible hands, patterns which now with the help of AI become more visible, and if so, to what extent can we intervene?

From a liberal point of view, it is of course important to hang on to the belief that we still have some freedom, that *to some extent* the future is in our hands. Contemporary liberalism is rooted in Enlightenment optimism and wants to but also believes it is *possible* to liberate and develop humanity, (re-)shape society and – by extension – shape the future of the planet. But this belief will come under pressure once we know more about patterns and correlations in the natural and social world, with the help of AI and big data. When predictions by AI become better, it might be more difficult to avoid fatalistic attitudes, also politically. Consider the ideal of democracy: what does 'self-rule' or even 'rule' mean if we feel, informed by AI, that we are in the hands of physical and social forces and that AI can predict the future? What kind of liberalism and democracy is then possible, if at all?

Consider again the Anthropocene idea. With regard to agency, the term can be interpreted as referring to a paradox: on the one hand, humanity has acquired hyper-agency, massive power over nature, but, on the other hand, once the new condition is in place, there are forces at play which we can no longer control, for example, human-influenced climate change. As humanity, we have never been so much in control; at the same time, we have never found ourselves in such a helpless and (it may seem) hopeless situation, one that is *out* of control. This is a paradox, since actually we *can* do something to change our predicament. But this 'something' is not the actions proposed by technological solutionism (which only makes things worse by increasing our grip on the earth) nor by fatalism – doing nothing because we believe that we cannot change anything anyway. To get out of the problem we need to radically transform our relation to nature and to one another. This also means our relation to ourselves as humans and humanity. If we need a humanism at all, it is one that changes the very meaning of what it means to be human, perhaps in a more ecological direction. One could argue that we need a different humanism, one that somehow is able to deal with climate change and AI, and does not cause the current problems. The answer to the problem of the Anthropocene is then: give us a different 'Anthropos', reconfigure the human-nature relation and thereby the human.

But humanism has also come under threat. AI and data science are themselves a challenge to humanism or are even likely to destroy it, as Harari (2015) has argued. Will we soon believe only in (big) data instead of the human? The technological and scientific developments around AI also seem to question human freedom in a political sense, but also in a metaphysical sense: do we still have free will (Harari 2018)? This is a long-standing discussion in modern philosophy since at least Kant: Do the (natural) sciences render free will problematic? I limit my discussion to the problem of *political* freedom. But as I suggested in the previous paragraphs, the issue concerning determinism is at least politically *relevant*. For example, the belief that we cannot do anything about our predicament may prevent rebellion against the climate-capitalist system, or may make it easier for people to accept totalitarianism.

Yet there is also another challenge with regard to humanism, one that is more directly relevant to freedom as a political problem: humanism's very focus on humans is problematic. What is largely missing so far in the discussion I presented is political consideration for non-humans and the environment. We should consider the idea that non-humans and the environment could somehow be politically valuable in themselves. The way the climate crisis has been defined so far and the various responses I explored based on mainstream political philosophy and political theory tend to assume that the 'who' which counts, morally and politically, are human beings. It is *their* survival that is at stake. It is *their* vulnerabilities that count; *they* are the ones at risk. It is all about *their* liberty, or it is all about justice and equality for *them*. It is all about *their* social order. But what happens if we change that assumption and include non-humans and nature into our normative frameworks for thinking about freedom and the politics of climate vulnerabilities and climate justice? What if we reject the assumption operative in discussions about the 'Anthropocene' that the problem is all about humans? What if we include animals in political history? What if we radically redefine the political collective and the social order in a way that includes non-humans and moves beyond the human-nature binary? This is the topic of the next chapter.

References

Arneson, Richard. 2013. "Egalitarianism." *Stanford Encyclopedia of Philosophy*, Summer 2013 Edition. Edited by Edward N. Zalta. Accessed 24 March 2020. https://plato.stanford.edu/entries/egalitarianism/#Pri

Coeckelbergh, Mark. 2013. *Human Being @ Risk: Enhancement, Technology, and the Evaluation of Vulnerability Transformations.* Cham: Springer.

Coeckelbergh, Mark. 2015. *Environmental Skill: Motivation, Knowledge, and the Possibility of a Non-Romantic Environmental Ethics.* New York: Routledge.

Coeckelbergh, Mark. 2017. "Beyond "Nature": Towards More Engaged and Care-Full Ways of Relating to the Environment." In *Routledge Handbook of Environmental Anthropology*, edited by Helen Kopnina and Eleanor Shoreman-Ouimet, 105–116. Abingdon: Routledge.

Crutzen Paul J. 2006. The "Anthropocene". In *Earth System Science in the Anthropocene*, edited by Eckart Ehlers and Thomas Krafft, 13–18. Berlin: Springer.

Crutzen, Paul J., and Christian Schwägerl. 2011. "Living in the Anthropocene: Toward a New Global Ethos." *Yale Environment 360*, 24 January 2011. Accessed 2 March 2020. https://e360.yale.edu/features/living_in_the_anthropocene_toward_a_new_global_ethos

Di Paola, Marcello. 2018. "Virtue." In *The Encyclopedia of the Anthropocene*, vol. 4, edited by Dominick A. DellaSala and Michael I. Goldstein, 119–126. Oxford: Elsevier.

Foucault, Michel. 1980. *Power/Knowledge: Selected Interviews and Other Writings, 1972–1977.* Translated by Colin Gordon and Leo Marshall. Edited by Colin Gordon. New York: Pantheon Books.

Foucault, Michel. 1988. *Technologies of the Self: A Seminar with Michel Foucault.* Edited by Luther H. Martin, Huck Gutman and Patrick H. Hutton. Amherst: The University of Massachusetts Press.

Foucault, Michel. 1998. *The History of Sexuality Vol. 1: The Will to Knowledge.* Translated by Robert Hurley. London: Penguin Books.

Frankfurt, Harry. 1987. "Equality as a Moral Ideal." *Ethics* 98 (1): 21–43.

Gabriels, Katleen and Mark Coeckelbergh. 2019 "'Technologies of the Self and Other': How Self-Tracking Technologies Also Shape the Other." *Journal of Information, Communication, and Ethics in Society* 17 (2): 119–127.

Harari, Yuval Noah. 2015. *Homo Deus: A Brief History of Tomorrow.* London: Harvill Secker.

Harari, Yuval Noah. 2018. "The Myth of Freedom." *Guardian*, 14 September 2018. Accessed 17 March 2020. https://www.theguardian.com/books/2018/sep/14/yuval-noah-harari-the-new-threat-to-liberal-democracy

Harrison, Peter. 2011. "Adam Smith and the History of the Invisible Hand." *Journal of the History of Ideas* 72 (1): 29–49.

Hayek, Friedrich A. 1960. *The Constitution of Liberty.* Chicago, IL: University of Chicago Press.

Heidegger, Martin. 1966. *Discourse on Thinking*. Translated by John M. Anderson and E. Hans Freund. New York: Harper & Row.

Marcuse, Herbert. 1964. *One-Dimensional Man: Studies in the Ideology of Advanced Industrial Society*. London: Routledge, 2007.

Marx, Karl and Friedrich Engels. (1848) 1992. *The Communist Manifesto*. Translated by Samuel Moore. Edited by David McLellan. Oxford: Oxford University Press.

Nussbaum, Martha C. 2007. *Frontiers of Justice: Disability, Nationality, and Species Membership*. Cambridge, MA: Harvard University Press.

Page, Edward A. 2007. "Justice between Generations: Investigating a Sufficitarian Approach." *Journal of Global Ethics* 3 (1): 3–20.

Rawls, John. 1971. *A Theory of Justice*. Cambridge, MA: Harvard University Press.

Sen, Amartya. 2004. "Why We Should Preserve the Spotted Owl." *London Review of Books* 26 (3): 1–5.

Smith, Adam. (1759) 2009. *The Theory of Moral Sentiments*. New York: Penguin Books.

Van Den Eede, Yoni. 2019. *The Beauty of Detours: A Batesonian Philosophy of Technology*. Albany, NY: SUNY Press.

Winner, Langdon. 2017. "Rebranding the Anthropocene: A Rectification of Names." *Techné: Research in Philosophy and Technology* 21 (2–3): 282–294.

Figure 6 Bear Glacier, Alaska, USA.

6 Enlarging the Collective

Toward a Political Science and a Less Anthropocentric Conception of the Body Politic

Introduction

In the previous chapters, it was assumed that the politics of climate change and AI, including the problem of political freedom, is all about humans. But as we will see, some argue that politics is also about non-human animals and other living beings. What about *their* freedom and political status? Some would even include non-living things. And the very things called 'climate' and 'AI' are as much about humans as they are related to non-human nature and agency. Furthermore, the term 'Anthropocene' is by definition very anthropocentric; can we move toward a less human-centered thinking? What happens with our thinking about freedom, climate, and AI when we cross these borders?

Using contemporary political theory (especially Donaldson and Kymlicka) including posthumanist theory (Haraway and Latour), in this chapter I explore what it means to think about the politics of climate change and AI in a way that (a) expands the borders of the political to include non-humans, (b) recognizes the political character of science, and (c) conceives of nature in a non-romantic way. This critique of some of the basic notions we have worked with so far (politics, the collective, science and technology, nature) is not only important in its own right, for example, in giving due respect to the political status of animals and ecosystems. A less anthropocentric conception of politics and political freedom also reflects back on humans: it helps us to further articulate a more relational view of humans and their freedom. I also refer to another political theorist, Alasdair MacIntyre, to support this relational view (which loops back to the connection I made between freedom and virtue), emphasize becoming as a central characteristic of human being, freedom, ethics, and the 'body politic,' and revisit some of the themes of the previous chapters such as Leviathan and Anthropocene in the light of these reconfigurations of politics, science, and the human-environment relationship.

Let's start with animals.

Citizenship for Non-human Animals

Aristotle famously claimed in *Politics* that humans are by nature political animals: 'man is by nature a political animal' (1253a). He did not mean that humans are natural 'politicians' (in the contemporary sense of the word) but rather that humans need others to live and flourish. According to Aristotle, only beasts or gods are sufficient for themselves and can live in isolation; humans have a social instinct. Therefore, they live together in various kinds of associations. These are associations of living beings with a sense of good and evil, just and unjust, etc. (1253a) – in other words, a sense of ethics. The *polis* (city state), then, is the association that, according to Aristotle, enables humans to achieve the highest good, human flourishing. But what about non-human animals? Aristotle excludes them from the political. Non-human animals lack speech and rationality. But surely some animals also live in 'communities' (Abbate 2016) and not in isolation, so why are they not political? Don't animals have needs, interests, and capabilities? And are they always completely unrelated to us? Are we not associated with them, too?

In their book *Zoopolis* (2011), Sue Donaldson and Will Kymlicka have argued for linking animal rights to political citizenship: animals can be considered as citizens. While animals, like children and some adults, have a reduced capacity for exercising political agency, this does not mean that they are not citizens. Thus, in contrast to Peter Singer (1975) and other animal rights theorists, for Donaldson and Kymlicka it is not the animal's intrinsic worth that makes them count but rather their citizenship (albeit a reduced form) and, more generally, human-animal relations. The various ways they are related to us implies that we owe them something, politically. Animals do not only have negative rights (don't exploit them, don't interfere with them, etc.), in the same way as there are negative, non-relational human rights grounded in human personhood. Since animals are citizens and are related to us in various ways, humans also have 'positive relational duties' toward them. For example, we should respect their habitat and care for animals who have become dependent upon us. This does not mean that all animals should be treated the same. We have all kinds of different relations with them, and just as human citizenship has many forms and includes many relations, there are also different forms of animal citizenship and corresponding duties. For example, according to Donaldson and Kymlicka, domesticated animals are full members of our political communities, whereas others are 'denizens' (second class citizens) or have their own communities – just like in present human societies.

This account does not reject arguments made by animal rights theorists in ethics: it acknowledges that animals have inviolable rights in virtue of being sentient beings (and thus buys Singer's argument). But it adds to this a framework for positive relational duties grounded in animals' political citizenship and the many relations we have with them. Similarly relational, but in contrast to Donaldson and Kymlycka (and to my book *Growing Moral Relations*) more limited in scope since responding to one particular political theory, contractarianism, I have argued that animals could get rights on the basis of relations of cooperation we have with animals, on the basis of their being part of a cooperative scheme – an argument which applies especially to domestic animals (Coeckelbergh 2009). Donaldson and Kymlycka also think that we have specific duties toward domestic animals. Insofar as these political rights are only based on *existing* relations, however, these arguments may be too conservative. We can also change relations and change the political order. Note also that, in the meantime, there have been further discussions of political rights for animals. Consider, for example, Robert Garner's *A Theory of Justice for Animals* (2013), which invites us to consider the question of justice for animals. It discusses Rawls but also responds to both Kymlicka and Coeckelbergh. And drawing on Joseph Raz's account of interest-based rights (instead of the inherent value of animals), Alasdair Cochrane has argued that animals should not be excluded from justice; however, and in contrast to animal liberation theorists, he thinks they do not have a right to freedom (Cochrane 2012).

Regardless of disagreements between these theories, these are novel and interesting views and approaches, since animals have so far been largely excluded, perhaps not from moral consideration, but surely from *political* consideration. Politics has always been defined in human terms. Since at least Aristotle, humans are the political agents and patients. Donaldson and Kymlicka question this. Not only are humans political animals; other animals are political too, in the various ways outlined here. They relate to us in various ways, they have interests, and the question of justice also applies to them. And in the light of the discussion in this book, we could add: in so far as animals are part of the political community (or political communities), the question of *freedom* also applies to them. I will return to this issue.

Opening up the Collective to Non-humans

Latour's Political Ecology

From another side, and with an interest in building bridges between science and politics, Latour has also proposed to expand the political: he argued for including 'things' in the collective (Latour 1993) and proposed that we talk about a 'politics of nature' (Latour 2004). Here the non-humans that were previously excluded from politics are not animals but scientific instruments and 'things' such as global warming, which are natural and political at the same time. This requires some explanation.

In *We Have Never Been Modern* (1993), Latour argued that during the modern period there was a separation between science and politics, or at least people had the idea that there is a separation. But in practice things get mixed and merge, and so it has always been. To take up a climate change-related example: the hole in the ozone layer is not just a chemical issue but also a political one. An epidemic is about biology but also about society (Latour 1993, 1–2). As I am writing, there are a lot of political discussions about climate change and about the virus COVID-19, also known as the Coronavirus. Both things/issues are scientific and political at the same time. They are what Latour calls 'hybrids'. (And so are the metaphors used in our discussion so far: Leviathan is a constitution, snake, and automaton; the invisible hand refers to a biological thing but also a political and economic force.)

According to Latour, moderns try to purify hybrids and separate politics and science, but this has never been successful. In this sense, we have never been modern. Just like in the pre-modern cultures Western anthropologists have studied, the political collective consists of hybrids. This has implications for thinking about democracy. At the end of the book, Latour proposes 'the Parliament of Things' (144). In such a parliament, natures are represented by scientists, and societies are present with objects such as the ozone hole, next to representatives of workers, industry, etc. Something like the hole in the ozone layer or a particular virus that people talk about are 'object-discourse-nature-society' (144). They are hybrid human/non-human. In the Parliament of Things, humans and non-humans are gathered in one political space according to a 'new Constitution' (144–145), which performatively produces new hybrids.

In *Politics of Nature* (2004), Latour continues his conceptual work to 'bring the sciences into democracy', as the subtitle says. How to include non-humans into politics? He argues that 'politics does not fall neatly on one side of a divide and nature on the other' but has always been related to nature (1). But the point is not that we have to focus on 'nature' and again assume that this is entirely different from politics; this would be too dualist and modern again. Instead of distinguishing between questions of nature and questions of politics, Latour proposes a politics of nature in the form of a *political ecology*. Instead of thinking in terms of nature versus society, or in terms of an assembly of things versus an assembly of humans, and indeed going beyond the (traditional conception of) 'nature' itself, he defines a 'collective' consisting of human and non-human members. Again, the point is that politics and science, society and nature, are assembled and mixed. And again Latour gives an example that is relevant to our topic of climate change: the Kyoto Protocol, adopted on 11 December 1997 and aimed to reduce greenhouse gases emissions in accordance with agreed targets, did not consist of two assemblies – one for politicians and one for scientists – but just one, bringing together political figures such as politicians and lobbyists but also scientists and researchers. This reflects that politically speaking there is 'a single collective' (Latour 2004, 56).

Like his notion 'the Parliament of Things', Latour's concept of political ecology aims to capture that the political world is composed of subjects and objects, humans and non-humans. He aims to arrive at 'a political ecology of collectives consisting of humans and non-humans' (61). In forums such as Kyoto, new entities participate in collective life (67). There is a fusion of different forms of speech. Or to take up a contemporary example: things such as viruses cannot speak on their own, of course, but they can speak through humans. Democracy, for Latour, means to also hear the voices of non-humans:

> To limit the discussion to humans, their interests, their subjectivities, and their rights, will appear as strange a few years from now as having denied the right to vote of slaves, poor people, or women. (…) Half of public life is found in laboratories; that is where we have to look for it.
>
> (69)

Thus, using Latour we can say that the Coronavirus is not something that belongs only to the sphere of the laboratory or the academic hospital, narrowly defined as belonging only to science; it is part of the collective and of the public life. It is what Latour calls an *actant*, a non-human actor. And, like any actor, it is a troublemaker (81). It stirs things up. Or put more positively: the laboratories nourish the expanding collective: public life collects 'the world that we hold in common' (91). This world is not fixed or given; it has to be performatively defined and discussed, and it changes. A new virus is now included in the collective and it is both discussed and researched; it is not just something that (to use Latour's word) the 'lab coats' study but is part of our public discussions and part of our common world. This is possible since health and epidemiology were always already public and political. In Latour's language: 'humans and nonhumans … find themselves gathered, collected, or composed' in the whole (131).

With this non-modern approach, Latour radically questions the nature-Leviathan binary, the modern idea that there is and should be a divide between a totally natural edifice and 'a totally artificial being', Leviathan (183). Seen from a non-modern perspective, politicians and scientists no longer belong to separate houses but use their skills to work on the (one) collective: 'to stir up the collective as a whole and get it moving' (203). Political philosophy, then, is not just about subjects or humans; it also about objects and non-humans, about 'administering the sky, the climate, the sea, viruses, or wild animals' (204). Using the term ecology, Latour welcomes non-humans into politics (226) and asks us to build a common world. For this we may need new 'institutions of public life' (228). In this sense, Latour concludes, 'the Leviathan remains to be constructed' (235).

Haraway's Posthumanist Recompositions

From a posthumanist angle, Haraway has also proposed to rethink and redefine the boundaries of the political collective. In one of her articles, she has argued that the term Anthropocene is far too anthropocentric (Haraway 2015). Not only *humans* have transformed the earth: there have been many 'terraformers' such as bacteria (and we should add: viruses!) and there are many interactions between different kinds of species and technologies. Our species, even if its processes have had planetary effects, does not act alone. There are also other 'biotic' and 'abiotic' forces (Haraway 2015, 159). Haraway thinks politics should aim for 'flourishing for rich multispecies assemblages that include people': the human but also 'the more-than-human' and 'other-than-human' (160). In line with her previous work on cyborgs (Haraway 2000), her concepts fuse the biological, cultural, political, and technological. In her posthumanist – or, as she puts it, 'composist' (Haraway 2015, 161) – understanding of the political, politics is not just about humans but about assemblages and (re)compositions, of which humans are a part. Humans are not the only ones around, politically speaking. They share the political with many other entities. The result is again a hybrid body politic. Here the political monster that remains to be constructed is not a totalitarian (and totalizing) anthropocentric Leviathan, but 'the diverse earth-wide tentacular powers and forces and collected things with names like Naga, Gaia, Tangaroa (burst from water-full Papa), Terra, Haniyasu-hime, Spider Woman, Pachamama, Oya, Gorgo, Raven, A'akuluujjusi, and many many more' (160). Here the political collective is inclusive and welcomes all kinds of human and non-human refugees. Haraway's body politic is also deeply relational and is multiple: it has many parts, tentacles, and connections. Similarly, in *Facing Gaia* (2017) and related recent work (e.g. Latour and Lenton 2019), which further develops his project of political ecology, Latour argues that Gaia should not be interpreted as a single organism or New Age goddess (let alone as a machine or thermostat); rather, it has the character of multiplicity. It consists of many agents that modify the earth. (If, as Latour shows, the term 'Gaia' is so confusing, however, one may also choose – contra Latour – to *not use it in the first place*. And Haraway's collection of names at least sounds New Age. Yet I appreciate how difficult it is to find a good new term.)

In any case, while probably not going far enough, Haraway's reconceptualization of the body politic offers an interesting relational way of thinking that responds to the time of crisis in which we find ourselves. In her book *Staying with the Trouble* (2016), Haraway has continued thinking along these lines, arguing for a 'tentacular' thinking. For her this is necessary when all beings of the earth live in disturbing times and when our task is to respond to devastating events. She says that we live not in the Anthropocene but in the Chtulucene, a word that names 'a timeplace for learning to stay with the trouble of living and dying in response-ability on a damaged earth'. In reply to the 'dictates of both Anthropos and Capital', her political imagination evokes monsters 'replete with tentacles' that live in 'multicritter humus but have no truck with sky-gazing Homo' (Haraway 2016, 2). Here the collective becomes a hot compost pile, an entanglement of mortal earthlings 'in thick copresence' (4). Haraway asks us to cultivate multispecies flourishing and multispecies justice and argues against the 'comic faith in technofixes' (3) and against despair and indifference (4). Her tentacular thinking, influenced by Latour, Arendt, and others, is aimed at 'sympoiesis': 'making with' (5). Instead of self-organizing (autopoiesis), which suggests that things make themselves, we make with others (58). This requires feeling and attachment (tentacles means 'feelers' – 31) and is all about relations: relations between the living and even relations of the living and the dead (8). We are never alone. In this way Haraway goes far beyond human exceptionalism and Western individualism. She proposes to 'chop and shred human as Homo' and rather make compost (32). It becomes increasingly impossible to still see ourselves as part of human-only histories (30–31). And for the same reason it becomes increasingly impossible to define politics in human terms alone.

Each in their own way, Latour and Haraway make the 'body politic' more inclusive and radically expand and transform the notion of politics. This is interesting for thinking about 'things' such as *climate change* and *AI*, and indeed for thinking about *freedom*.

Implications for Thinking about the Politics of Climate Change and AI

What are the implications of these ideas for thinking about the politics of climate change and AI, including the problem of *freedom* that is central in this book? Let me articulate some implications, with a focus on using Latour:

First, Latour's view supports the main idea of this book that we should understand and discuss climate, AI, and the relation between the two *as a political problem*. To deal with both climate change and the challenges raised by AI, we better recognize how political 'climate' is and how political AI is. Climate change is not just a scientific issue but also a political one. And AI, too, is not just about science and technology; it is also about power and politics. Both topics are not and should not be discussed in two different assemblies but in one. Both topics have a hybrid human/non-human character. They should not be banned to a separate natural or technological realm; they are deeply political and human. Furthermore, with Haraway and taking seriously the political dimension of narrativity, we could add that narratives about AI and about climate change are also part of the political. Her cyborgs and assemblages cut across the fiction/non-fiction boundary; AI and climate change are also such hybrid monsters. The political is about what actually happens to, and between, people and things, but also about the stories we tell each other.

Second, Donaldson and Kymlicka, Haraway, and Latour therefore invite us to think about politics and the political problems concerning climate change and AI in a way that *includes non-humans*: not only as topics but also as political entities, e.g. political citizens or members of the collective. This is quite a change. Traditional views of politics (modern and ancient) assume that politics is only about humans, and in particular their interests, voice, and language. But Donaldson and Kymlicka add animals to the body politic. We could conclude from their view that the problem of climate change as a political problem also concerns animals, to whom we may have duties of justice and compassion. Similarly, if AI is a political problem, then adopting a more inclusive understanding of politics implies that we should also consider how AI impacts animals and ecosystems. Going further, Latour asks us to welcome all kinds of non-humans such as things, facts, etc., into the collective. Applied to the problem at hand: politics will be increasingly (and always was also) about non-humans such as climate and AI. The climate and AI have become part of the public life and are brought into the collective in various ways by politicians and scientists, each using their own skills to try to move things. For example, political activists and politicians *speak for/in the name of* the climate. The point is not that these non-humans have necessarily intrinsic value or are in any way human-like. Maybe they have and are, maybe not. These arguments are not about their moral (intrinsic) value or status, but about their *political* agency and patiency. AI and climate change are part of the political collective and the public life, and in this sense it is recognized that they have interests, they are given a voice through representatives, they appeal to the public, and so on. In order to deal with climate change and AI, we have to better understand the role of non-humans in the social order and the new political challenges that arise from this. These views redefine this role and the related challenges.

Adopting a Latourian way of thinking about politics implies, for instance, that we should discuss the role AI plays in politics, who speaks for it, what place we give it in the collective, etc. If AI is part of the collective, and if this collective takes the form of a representative democracy (rather than Rousseau-type self-rule) as Latour assumes, then we can say that AI is represented not only by scientific experts but also by representatives who are traditionally seen as 'political' such as lobbyists.

And, indeed, this is what we see in advisory bodies on AI, for instance, which are often hybrid Latourian groups made up of experts and people representing interests. Similarly, from a Latourian perspective, it should not surprise us that climate change is discussed by scientists but also by all kinds of political actors, including, for example, teenagers, who play all kinds of representative roles and speak on behalf of the climate, future generations, and other humans and non-humans. Non-humans are represented in various ways. For example, Greta Thunberg speaks for the facts of science, for the climate, for the future of the planet, for her generation, and for future generations. And not only humans but also all kinds of animal species and coral reefs are part of the political discussion about climate change. Even AI and data science join the club. Climate forums, social media, and (other) public discussions assemble all these hybrids and representatives. The collective that is gathered around climate change becomes, in Haraway's words, a hot compost pile. (And a similar interpretation can be offered with regard to the Corona crisis discussion, understood as a non-modern assemblage or compost containing viruses, ventilators, masks, numbers, graphs, health care workers, scientists, politicians, patients, and so on.) Climate change and AI show how non-modern politics has become, and probably always has been. Once we adopt a posthumanist and non-modern way of looking at politics, we see that modern categories no longer suffice to describe what is going on.

Third, if Latour is right that politicians and scientists are part of the same political collective and work on the same chain of humans and non-humans, and if Haraway's project to blend narrative and technological practices makes sense, then it seems to me that education cannot suffice with modern monodisciplinarity: even if each kind of (human) political participants might still gain some specific skills and knowledge, it will become increasingly important that they can talk to one another and engage in *common* political-scientific projects in order to deal with the hybrid nature of problems such as climate change and AI. This requires the learning of each other's languages and the development of interdisciplinary and transdisciplinary methods and tools that cross and connect not only academic disciplines but also different worlds such as science and politics. For example, if Coronavirus experts talk to politicians who have no idea about science at all, things go wrong. If AI experts and data scientists cannot explain what they do to (other) political interest groups, there is little chance that global challenges with regard to AI can be solved. And if citizens are not educated about climate change, they may not be able to adequately defend their own interests (let alone the common good) in and through the relevant political institutions. Building the collective in the form of a non-modern community requires *communication*. The expansion of the collective Latour talks about, and indeed the building of a common world – including the creation of a new

Leviathan, that is, new political institutions that can deal with issues such as climate change and AI – requires a *radical transdisciplinarity* that enables both scientists and non-scientists to communicate and bridge the divide(s) modernity tried to create. This is part of the kind of expertise and wisdom we will need in the politics of the 21st century. If we need Aristotelian *phronesis* at all, then this practical rationality needs to take the form of a transdisciplinary, border-crossing kind of communicative rationality. With Latour we can say that we need collaboration between scientists, activists, artists, etc. However, one could also go further and say that we do not only need to strengthen the communication between experts on humans and experts on non-humans, between the humanities and the natural sciences, since that assumes both are first educated and practiced in their own silos and are then asked to talk to one another (as is the case now in interdisciplinarity). Instead, we need to go further and build a transdisciplinary foundation that better connects both kinds of fields *from the very beginning*, that is, one that makes connections between different worlds and recognizes the hybridity of problems *already within the knowledge practices and educational practices and institutions* at all levels, from kindergarten to university. We need people that can deal with hybrids. We need people that can help to shape the new climate regime. We need people that know how to build a hot compost pile. For this purpose, we need to merge both educational trajectories to some extent at least.

Fourth, for the question concerning political *freedom* these views imply that, normatively and politically, not only does the freedom of humans count, and maybe even that freedom should not only be defined in human terms. What about the negative and *positive* freedom of non-human animals? What about *their* needs and capabilities? What kind of freedom do *they* need? If there really are good reasons to include them in the body politic (because they are sentient, perhaps, but also and especially because they have interests, because we relate to them, because we cooperate with them, because they have capabilities, etc.), then when we think about the future of our planet and the use of AI, we need to take them into account, politically speaking. This means, among other things, thinking about their freedom. For example, AI and data science might not only be a threat to human freedom, for example, by taking mass surveillance to a new level and enabling authoritarian rule; they also may increase the power we have over non-human animals and may lead to more restrictions to *their* negative freedom. And climate change may lead to less opportunities for animals to realize their capabilities and hence jeopardize *their* positive freedom, potentially leading to death and extinction.

Moreover, thinking about the freedom of animals also helps our discussion about political freedom for *humans* in the following ways:

First, we can better understand the limitations of having only negative freedom and the importance of positive freedom for *humans* if we consider the freedom some animals have today. Consider the freedom of fish whose marine environment changes rapidly as a result of climate change and who can no longer find their food, or the freedom of a koala bear in a burned down forest (which, we shall assume, is partly and indirectly the result of climate change). The fish is free in a negative sense, but can no longer survive. The koala is free, has all negative freedom there is but has nowhere to go for food and will die without external help. Capability freedom or positive freedom is entirely lacking, because its habitat is destroyed. Now the fish or the koala bear cannot help that this happened; humans, by contrast and tragically, are destroying their own habitats, the habitats on which they depend. In other word, humans are destroying their own positive freedom in the name of negative freedom – while destroying the positive freedom of other animals.

Second, discussing the freedom of animals helps us to better understand how *human* positive freedom is also linked to biology and ecology. For animals, survival, but also their positive freedom, flourishing, and 'the good life', depends on the needs and capabilities they have as a particular lifeform and species. We readily accept this in the case of non-human animals. But it is also true for human animals. Maybe human capabilities are not 'fixed'. Perhaps we have expanded the sphere of good(s) in the course of our history as a species. What is good for humans has been expanded. As has been argued in philosophical anthropology, we are not 'finished' animals. We are not yet 'fixed' by nature, as Nietzsche put it in *Beyond Good and Evil.* We have altered the human and we are altering it. We are always becoming. Technologies and imagination change the human and expand our goods. By means of our imagination we create new needs and desires, and our possibilities can be extended/enhanced by means of technology (which in turn creates new needs and desires). But any good is still related to, and dependent on, the possibilities of our body and the affordances of our natural environment and, in the end, the earth. This puts limits to the possibilities we have as humans. Even if 'enhanced' by technology (e.g. AI) and even if we create new capabilities, we remain dependent on our body and our natural environment. The good life for all animals, including humans, is intrinsically connected with the good of our environment and the health of our body, which itself is always related to the natural environment, including weather and climate. Any ecological understanding of the human links the question about human good (and hence also positive freedom) to the question about the environment and the climate. Positive freedom cannot be defined outside any environmental or ecological context. To put it stronger: it presupposes such a context. (I will return to this when reviewing MacIntyre later in this chapter.)

Furthermore, next to freedom, there are *other political values and principles* that are important and other concepts that can and should be used when discussing the politics concerning both human and non-human animals: needs, capabilities, survival, justice, equality, and so on.

As in the case of humans, politics concerning animals can and should not just be about freedom. For example, Donaldson and Kymlicka think that politics is also about justice and compassion. Based on this view and as I already suggested, one could argue that if we discuss climate change, we should ask not only about justice for humans but also about justice for non-human animals. The same is true for AI and its impact: politics of AI should not only be about the impact on humans and human societies; it should be about the impact on entire ecosystems, including animals, who are part of the political collective. And this impact is not only a matter of freedom; there are also other political values at stake, such as, for example, justice or equality. Is it right, for instance, that we treat some animals better than others, including better protect some animals from the impact of climate change and AI than others, and so on?

Further Questions and Directions: Beyond Nature and the Politics of Becoming

In What Sense Are AI and Climate Change Political?

Some readers may think that the posthumanist claim that AI or climate is political means that AI or climate have moral standing, perhaps even a moral standing like the moral standing of a person. But this does not follow. Similarly, Latourian *actants* that play a role in the debate about AI and climate change are also not necessarily political agents or patients in the way humans are. Let me critically discuss how and in what sense AI and the climate are or can be political.

First, following Latour, *AI itself* can be part of the political collective. But can it also be a political agent and if so, in what sense? What does it mean to say that AI 'does' things or 'acts' in the political sphere? And can it be a political patient? Can and should it be the object of political concern? Should its voice be heard? Should it be given a voice and be represented in our assemblies? What does 'freedom' mean for an AI? Perhaps AI can have a kind of negative freedom (or not): its actions may be restrained or not. But is that 'real' freedom, meaning is that the same kind of negative freedom that humans have? If not (e.g. because it is argued that only humans have free will whereas AI lacks this), what kind of freedom is it? Does it even make sense to speak of the freedom of a technology? Moreover, can AI have positive freedom? At first sight, the concept of positive freedom has a lot to do with human nature: it is defined in a way that presupposed human nature, including human needs, capabilities, and possibilities. Therefore, it seems, we cannot apply this concept of liberty to AI. Or can we define capabilities in a non-anthropocentric way?

One could argue that AI can 'act' politically in the sense that humans talk about it as if it has agency and 'does' things, but that it cannot be a true or full political agent or patient at all since, although it may have causal power and hence causal agency, it lacks *interests* in any common meaning of the word. It is not a subject and it is not aware of anything. It may have a goal: a goal that is programmed or a goal that it calculates on the basis of another kind of goal. But it doesn't really *care*. It does not care about the goal and it does not care about achieving the goal, even if it will try to achieve it. Under conditions of climate change, in the climate crisis, the only entities that care are those beings that are alive and conscious. In contrast to AI, some animals care, they have interests, they are therefore *political animals*. Human or not, their liberty is at stake. And perhaps justice matters to them in some sense. Not necessarily because they themselves can think about things such as justice or liberty. A cat or a chimpanzee cannot *think* about that, in the sense of having the concept. And they may not care about justice in the same way as human beings do. But whether or not they are free, justly treated, etc., matters to them; the behavior of 'higher' social animals shows that they care. Therefore, we humans can, and should, think about their freedom, about interspecies justice, and so on. In *The Case for Animal Rights* (1983), Regan uses the term 'subjects-of-a-life'. Animals such as adult mammals and several species of birds are subjects of experience and their lives matter to them. According to this view, many animals should be recognized as being subjects of life and experience, and therefore – one could infer – they may not only have moral standing (this is not a question I discuss here) but they are certainly political, if not as full agents (if that means being able to speak and reflect) then at least as patients. They can be freedom patients or justice patients. Freedom and justice concern them. Now this does not seem to apply to AI. In a Latourian, non-modern sense, however, even AI that is not sentient, not subject-of-a-life, etc., can be part of the political collective. It cannot speak itself but can speak through representatives who give it a 'voice', 'defend' it, etc. Scientists and technology experts can give AI a voice, can defend AI, for example, against accusations from those who speak in the name of humans and humanity. In this sense, AI plays an active role in politics and is part of politics. In this sense, AI is a political actor (Latour: *actant*).

Note, however, that this Latourian interpretation of AI assumes that there is something like 'an AI'. In practice, AI is a combination of technological elements such as software (algorithms), data, and infrastructures. AI has its own hybridity. It is also connected to humans. This renders it at least a little misleading to talk about the moral or political status of 'AI' or 'an AI', as if it (always) appears in the form of a separate entity – say a humanoid robot powered by AI. That being said, one could include all the different parts and phenomena of AI into a Latourian or Harawayian collective. They can all be part of the hybrid human/non-human network, assemblage, or compost pile, at least and especially in so far as we (humans) talk about them and in so far as they have political consequences. And of course there is causal agency of technology, for example, when a self-driving car 'kills' someone. But these meanings of 'political AI' do not come anywhere near giving AI moral or political status as agents or patients, in the same way as humans are political agents or patients.

Second, with regard to *climate change*, things seem different. 'Climate', 'earth', etc., are not only part of politics in a Latourian sense that they are part of the collective; they are also connected to living beings and ecosystems. Here we should consider arguments that give moral patiency to natural beings and ecosystems, since this could be politically relevant. For example, according to animal rights advocates, sentient beings are moral patients. We have duties toward them, as an example, we should not make them suffer. And some would also say that the earth itself is a moral patient. 'Patient' not in the sense that the planet is 'sick' (although this is also said in the debate on climate change and environment) but in the sense that morally something is owed to it by humans. Not because the earth is conscious or a subject-of-a-life but because the earth and its ecosystems have intrinsic value. Here the tradition of deep ecology is relevant, which recognizes the inherent worth of living beings, ecosystems, and the earth (or 'the land'). The natural world is seen as a balance of complex inter-relationships: organisms are interdependent within ecosystems. There is a natural order, which then gets disturbed by humans and environmental harm and destruction. Such a deep ecology view can then be used to justify doing something about climate change: not for the sake of humans, but for the sake of the earth. One may want to 'save the earth' and indeed 'liberate' it from humans, in the same way as one may want to liberate animals. Thus, here natural entities can be moral patients to which (or to whom?) we owe something, to which or to whom we have duties. And this moral status is politically relevant. Since they have this status, it is important to give them a voice by means of representation. Rivers and atmospheres cannot speak for themselves. Therefore, it could be argued, we have to defend them and give them a voice. And from a deep ecology point of view, the main task of the representative is to defend the natural entity against humans. Thus, here the point is not so much that animals and ecosystems are part of the political collective or pile in a Latourian or Harawayan sense, but that they in addition have standing as political patients that need representation. Because animals have real interests and ecosystems have moral value (intrinsic value), one could say that they have a stronger kind of political standing than, say, AI technology.

Toward a Non-romantic Politics of 'Nature'; Beyond the Idea of a State of 'Nature'

However, the argument from deep ecology tends to assume that there is something like an external 'nature' (with its climate) that we need to protect or retreat from, perhaps something that was first virgin and pristine and that then has been spoiled by human intervention and technology (e.g. AI). This is a problematic way of conceptualizing the human-nature relation. Against this romantic Garden of Eden kind of view (and in line with some philosophers of technology such as Don Ihde), it could be argued that we have always technologically interfered with nature and that we are natural beings. Humans are not something external to nature. Moreover, the world as we find it crosses the nature/human binary. This calls into question the very use of the term 'nature'. How helpful is it really, given these misleading connotations? For example, environmental thinker Vogel (2002) has called for an environmental philosophy that no longer employs the concept of nature, since he claims that 'the world we inhabit is always already one transformed by human practices' (Vogel 2002, abstract), which makes humans responsible for transforming the world. In my own work on this topic, inspired by Tim Ingold and Martin Heidegger, I have used notions of engagement and care to argue that we actively relate to nature and in nature (and we are nature) (Coeckelbergh 2015, 2017). Politically, that means that the green political project should not so much be about defending and protecting nature from the big bad humans or from 'humanity' (as if it is an external monster, an Anthropocenic hyper-agent or invisible hand) and trying to go back to a fixed original state of nature when everything was still fine, but rather that we have to fine-tune and shape our ongoing, close relation to nature. The question is, what kind of active relation should we have toward our environment (which may be hybrid natural/artificial)? What should we do, given that we are already natural beings and deeply connected with nature through our activities? If we need a politics of nature at all (i.e. if we still want to use the term), then it needs to be framed in a less distant way, for example, as a politics of engagement and a politics of care. On this basis, the use of AI could be criticized on the basis that it contributes to enforcing and sustaining a relation that is based on control and domination. Both the bad Anthropocenic invisible mega-agent who messes up the earth and the well-intended but highly problematic 'green Leviathan' that tries to monitor and repair things are then problematic – among other things – because they are still obsessed by control. And the imaginary still suggests a human-nature dualism. We should try to move beyond that. Now what is the role of AI then? In what way can AI help us to redefine the human-environment relation (I avoid the term 'nature' because that makes it sound as if it is a thing or agent) in a way that mitigates the climate change problems and helps us to deal with the 'Anthropocene'?

Again, this is a way of framing the problem which, like Haraway's and Latour's interventions, questions the very idea of the problem of the 'Anthropocene'. From this perspective, the political problem regarding climate change and AI is not a kind of battle between human technological agency and 'nature', but a question of tuning the human-environment relation, which is already technologically mediated and shaped, *and* which – taking into account what Latour and Haraway say – is already connected with many other entities. With Latour (1993, 2004, 2017), we should question the modern notion of 'nature'. Instead of using the term Anthropocene, we need other notions that reflect this. Similarly, in the light of this critique of 'nature', the political question is no longer about how to deal with the 'state of nature', as in Hobbes or Rousseau. Climate and what used to be called 'nature' are themselves *already* political. Latour uses the term 'new climatic regime' (Latour 2017), which combines a term usually used in the sphere of politics (regime) with a term that was supposed to belong to the domain of science. Non-modern thinking cuts through this politics/nature dualism. Once we no longer assume a natural/artificial duality, the political task is one that is natural and artificial at the same time: it is about emergence and it is about making. It is a kind of *poiesis*, as Haraway suggests. As Heidegger shows in 'The Question Concerning Technology', this term expresses that one makes things but also at the same time participates in the process of making (Heidegger 1977; Coeckelbergh 2018). In this sense, the political is a work of poetry, in which humans participate. We do not have full control. Latour and Haraway are right to point out that there is a much wider political collective. In *Facing Gaia*, Latour points out that there are many connections between human activities and the natural world and many agents, also non-human agents that shape the earth. Yet while we humans are not the only poets, we are certainly one of the most important ones, given that we have the capacity to do a lot of political-poetic work. Insofar as we have this capacity, the political is our *poetic responsibility*. This is a truth in the Anthropocene concept: the more work we are doing, the more we are responsible.

But not only does politics need to be redefined once we take on board post-nature, non-modern, and posthumanist relational thinking. We are also invited to revisit the question concerning *freedom* in the light of a more relational view and in critical dialogue with non-modern thinking.

Toward a Relational Kind of Freedom and Becoming

In *The Human Condition* (1958), Hannah Arendt, following Aristotle, imagined ancient politics and political freedom as a freedom from household matters and a freedom from the metabolism, the biological, the earthly. The slaves do the work and deal with that, whereas the aristocracy can have a political conversation about the *polis*. The men talk about important matters and the Greek hero seeks to become immortal; others are slaves: slaves to others but also slaves to their own desires, body, and mortality. Freedom and politics are thus defined in opposition to the sphere of work and necessity. Similarly, modern thinking has always been trying to separate the domain of freedom (where politics and morality are at home) and the domain of necessity (nature, the domain of science).

But as Latour and Lenton (2019) have argued, there is continuity between the two domains once we consider the earth as a self-regulating and at the same time interdependent system. There is a kind of freedom defined as self-rule, democracy, but one that at the same time involves interdependence and is much more inclusive than politics defined in human terms only:

> When humans look at Gaia, they do not encounter the inflexible domain of necessity but, strangely enough, what is largely a domain of freedom, where life forms have, in some extraordinary ways, made their own laws. (...) Conversely, any human trying to situate himself or herself as part or participating in this history can no longer be defined only as 'free' but, on the contrary, as being dependent on the same sort of intricate and intertwined events revealed by Gaia. More freedom in the domain of necessity is fully matched by more necessity in the domain of freedom.
>
> (Latour and Lenton 2019, 679)

This deconstruction of the freedom/necessity binary leads Latour and Lenton to expand Aristotle's notion of politics and democracy: democracy is 'composed of all the political animals' making their own law, not just humans. The *demos* now consists of humans and non-humans. Moreover, the kind of ancient politics described by Arendt is both impossible and undesirable. Politics will always have to do with the guts and the dirt of the earth. With the living and with the not yet or no longer living. We should stop excluding work (*poiesis*) and labor from politics. Furthermore, politics should not happen at the expense of women, slaves, non-human animals, nature, and whoever and whatever else has been constructed as the other/opposite/different of the male superior and self-sufficient 'free man'. Freedom for humans needs not only the support of human others, as became clear in our objections to Hobbes; it also depends crucially on the ('natural') environment. The adult kind of freedom we should be seeking, therefore, is not the freedom of security from others which we mistrust and whose authority we rebel against (negative freedom) or the freedom of exploiting others (ancient freedom in a slavery system), but a positive freedom, one which is not so much about self-mastery (see Berlin again) understood as self-sufficiency by Aristotle and Rousseau, but one that is instead deeply relational and (inter)active, since it is not a freedom 'in spite of' but one that develops and becomes as it *relies* on others and on the natural environment. If humans can be free at all, it can only be a freedom in and through dependence.

This claim can be supported by relying on resources from outside the Western tradition, but it can also be elaborated in critical dialogue with Aristotle. This leads us to the work of Scottish-American philosopher Alasdair MacIntyre, who is known for his work on virtue and communitarianism, but who also defends a relational view of the human and ethics. This relationality we can further radicalize, give a posthumanist twist, and connect to the environmental discussion and to the discussion about positive liberty and capabilities.

MacIntyre (1999) argued that we are dependent animals: rational animals, perhaps (from a Hobbesian, pessimistic point of view this can be questioned when we see what people do to the environment, the planet, and to one another – and indeed when we see how some people behave in times of crisis, for example, the Corona crisis), but in any case, *dependent* and vulnerable animals. To flourish (and let me add survive) we have to acknowledge this dependence. MacIntyre makes a connection with virtue ethics and the good life here: to reach the good life and have virtue always is a development that 'has as its starting point our initial animal condition' (MacIntyre 1999, x). We share a lot with non-human animals, or at least – according to him – with some species, and it is a central feature of human life that we are vulnerable and dependent animals. This is the basis that we develop the virtues. Living with others and participation in community helps us to acquire the virtue, wisdom and (in MacIntyre's Aristotelian view that stresses rationality) practical rationality (*phronesis*) we need to live in a good way, that is, to be ethical. He questions Aristotle's emphasis on masculine virtue and the superiority ideal of his *megalopsychos*, at least insofar as his notion of honor and (literally) 'greatness of soul' is too unrelational. According to MacIntyre, who uses the more vague translation 'the magnanimous man' (7), Aristotle failed to give due recognition to affliction and dependence. But these are central to the human condition (4). We are relational and dependent beings, and this fact is and should be significant for moral and political philosophy. MacIntyre thus understands rationality and virtue as related to our social dependency and animality. Staying close to Aristotle and the Western philosophical tradition, he does not reject the modern emphasis on individual autonomy as such, but argues that virtues related to individual autonomy need to be exercised together with 'the virtues of acknowledged dependence' (8) since they are the prerequisite for developing oneself as a rational animal and achieving human flourishing (9). Through social relationships and participation in a community, humans learn the importance of both independence and dependence, learn giving and receiving, and discover their own good through recognizing the common good. We learn the common good through shared activities (136). This common good is defined in terms of acknowledged dependence itself. It is also about acknowledging risk: our vulnerability and dependence may always increase. MacIntyre envisages a political society 'in which it is taken for granted that disability and dependence on others are something that all of us experience at certain times in our lives and this to unpredictable degrees' and that consequently this is not a 'special' interest but the interest of the whole society and one that is 'integral to their conception of their common good' (130). He thinks that neither the modern family nor the modern nation state can provide this; instead, he argues that only a local community (and not a national state that pretends to be a community) is the right kind of form of association for ethical development and human flourishing.

MacIntyre's communitarianism, which seems to focus on local communities, is too limited when it comes to dealing with the challenges posed by climate change and AI at a global level. We still need collective action and the right kind of political institutions at that level, next to the development of community. Moreover, it is not just individual-biological and social dependence that renders possible ethics, as MacIntyre argued; ultimately it is also our dependence on other, non-human beings, on ecosystems, and on the earth that enables this ethical growth. His anthropocentrism, borrowed from Aristotle, must be questioned. Based on the views developed in the previous sections, one could argue that the common good and the ethical-political community should not be restricted to humans and their communities. And animality is not just an 'initial' condition which then gets modified or built on, if that means we can leave it behind once we become cultural, ethical, and political beings; biological dependency and social vulnerability are always present throughout our lives and everything we do. And so is our dependency on the wider natural environment, of which we are part. Paradoxically, and connecting to what I have said earlier in this book about freedom, capabilities, and virtue, these dependencies on others *and* on the natural environment give us a freedom: not a negative freedom but a positive one. MacIntyre's view fits well with a view of positive liberty as 'freedom to' and capabilities: developing ourselves within relations of dependence (but as I proposed including dependence on *non-humans* as well), we gain in capabilities, we can fulfill our needs, and indeed we develop virtues and learn to live well with others, which paradoxically includes some capacity for independence. If we adopt this conception of liberty, communitarianism is not anti-liberal but fits with liberalism, at least with this stretched form of it. Moreover, MacIntyre also offers us a more developmental perspective. The realization of capabilities and the attainment of flourishing does not just happen to us. We have to work on it. It requires a life-long development that entails active relating to others and (I add) actively relating to our environment(s). One could say that this kind of relationality is not so much something that 'is' but rather a kind of *relationing*. We need a verb to express that it is a process and a development.

According to this view, then, human being, freedom, and ethics are not about ontology, not about 'what is'. Rather, they are a matter of *becoming* and have to be understood in terms of verbs: it is about doing, transforming, caring, and so on. There is no such thing as a 'state' of nature; there is no *stasis*. And there is no thing called 'nature'; we are always already related to it as we and our environment change. We also relate to nature in an active way. If there is a 'nature' at all, it is one that we transform and in turn are transformed by. This ongoing transformative process includes the use of technologies. Technologies can be seen as part of this becoming and can support it. Thus, a more developmental and relational perspective on freedom and ethics means that climate change is not so much an 'external' kind of 'thing' that is there and that we 'face' as a kind of alterity. Rather, as part of a dynamic natural environment climate change is deeply connected to us and also changes us: as persons, as communities, as societies, as humanity. As relational beings, we are part of a holistic system. Deep ecology got that right. But, in contrast to what deep environmental ethics seems to assume, the world is not a collective of things, in particular things with 'intrinsic value'. It is always in movement and moving. Similarly, freedom and the political collective should be understood in these terms.

Becoming: Moving the Body Politic

Perhaps the very word 'collective' does not do enough justice to the process character and is too dualistic. In order to conceptualize a politics that becomes, we neither want nor need a metaphysics of entities and not even of relations or networks. Even Latour, in spite of mentioning movement and temporality, is somehow stuck in a non-moving, rather than fixed, metaphysics. The terms assemblage or network too much suggest a given collection of things. But we can do better: we can insist that the political collective is a matter of *becomings*. It is something that happens and it is also something we do and work for. It is something we have to shape. Making the collective, understood as a becoming, is *poietic*. But as Haraway teaches us, this is not a matter of individual making/becoming: it also *sympoeietic*: a making-with. We have to (re-)make the political together. And together, we are always related to our environment. We should think beyond a dualism of humans versus world or humans versus things and other entities. As we actively relate to our environment, politics is a becoming; it emerges.

Climate change in the Anthropocene, then, is not a something 'out there' that threatens us, as if it were an object or an agent. It is something that goes on and in which we are totally involved and which is in us, very much like an illness. It goes on in the ecological whole which can become, or host, a body politic. This body is both social and natural at the same time. Yet while the ecological whole is not just made by humans but evolved and is the result of much non-human agency as well, the body politic is partly but mostly constructed by humans. Given that we have now acquired so much power and agency as humanity, also at a global level, it has become mainly our *poietic* project. We are not an all-powerful demiurge of the political. Haraway is right that other species and entities also take part in it. But we have a big role, since we are one of the main political poets. Now often this making of the body politic is done in a way that excludes others (human and non-human) and that builds Leviathan automata such as states and corporations. Mortal gods that rule us. But we are invited to acknowledge, embrace, and work toward an alternative body politic that is more relational and non-authoritarian. We better make work of that, given that things are moving. If the body politic is a moving body that is to a significant extent our *poietic* project, we carry a lot of responsibility for it.

On Mortal Gods and Mortal Humans

This leads us back to the problem we started from in this book, when we considered the challenges posed by the idea of a green Leviathan, which threatens freedom. How can we make sure that this is not a despotic machine? Is the only way to 'save the planet' that we construct monstrous mortal gods? Or can there be good mortal gods? Can we move beyond mortal gods altogether in our political thinking? Can we have a truly non-authoritarian solution? Does this require too much trust in human nature? Does it require a transformation of human nature? What is the role of AI and, more generally, technology in this? Are ancient recipes enough or do we need human enhancement, for the sake of saving the earth and saving humanity? Or does that create again new automata, new Leviathans, new machine snakes, new mortal gods that take power – for better or for worse? Will we make human beings into a kind of green robots that do exactly what the Leviathan tells them in order to be environmentally friendly? Is this freedom? Is this the good life? Is this the price we have to pay, or are there other solutions?

When asking these questions, we should not forget the temporal dimension, in particular the aspect of mortality. Following Hobbes, I said that these monsters and gods are mortal. The social and political constructions we make and the artificial beings we create are fleeting, temporary, transient. Maybe democracy is the most vulnerable monster. It is not a Leviathan. It is not all-powerful. It can be easily destroyed. Yet all forms of government, also the most authoritarian and totalitarian ones, will once perish. All these forms of the political depend on people and ultimately on the earth and its ecosystems. Leviathan monsters are also part of the pile.

AI can become a new Leviathan. It can become part of a Hobbesian Leviathan: a monster that rules with the sword. It can take away human freedom. But it could also create better conditions for freedom and self-rule. By making transparent historical and present vulnerabilities and relations in the body politic, it could contribute to social and political change, survival, and justice. If a Leviathan is needed at all, if we still need a mortal god at all (and if therefore political philosophy has to remain a form of half secularized political theology), we should build a democratic Leviathan: new political institutions that are more inclusive and participatory – at national and global levels – and that support freedom as self-rule and development of capabilities. And, hopefully, this democratic but sufficiently powerful monster will offer guarantees against attempts to take away too much negative freedom and lead to more positive freedom, realization (and perhaps enhancement) of capabilities, flourishing, and the good life: for humans and for non-humans.

But even better would be not to build a political god at all, and rather help ourselves with political artifacts and political infrastructures that are democratic without being all-powerful and all-knowing. In such political automata, networks, and compost piles, power would be more diffuse and distributed. However, then there is still a problem of coordination, at local and global levels. Inclusive assemblages, networks, and hot piles may sound good, but it is not clear how they can deal with global problems. One could therefore argue that a political monster à la Haraway's Gaia or Pachamama would be needed to coordinate dealing with climate change at a global level. But replacing the patriarchical monster with a matriarchical one does not seem to solve the problem: it is still a monster, albeit a post-Hobbesian one. With Latour, we may want to reject being ruled by a monster or god once and for all. On the other hand, given that the earth is already in the state it is, it seems that we need some kind of planetary management, coordination, ruling, and governing of the planetary household (a *nomos* of the *oikos*) or ship – even if that very idea is also problematic, as we have seen. And what if another species than humans, for example, viruses or bacteria, takes over the world? Or indeed AI? Gaia is not just peaceful. There are political tensions; there is the risk of war. The common world and political collective we humans want is *one that includes us*; even if we are sympathetic to posthumanist political imaginaries and other theory that questions anthropocentrism, we probably want to retain at least some weak form of anthropocentrism along these lines. A loose assemblage or chaotic compost pile may not be enough to guarantee our inclusion. Haraway's Spider Woman or Latour's newly constructed Leviathan might do a better job to build that common world in the age of climate change, involving some centralized power and ensuring that humans, not against but *with* other lifeforms and technological entities, can survive and flourish. On the other hand, in so far as this solution entails centralization and authority, it raises again the classical political philosophical questions regarding freedom that we discussed earlier in this book. These are not Hobbesian Leviathans, but Leviathans nevertheless. This raises similar questions in light of new technological possibilities.

An AI-powered monster or superheroine has many tentacles and powers, and freedom (of humans and non-humans) may easily become their fragile prey. It remains a huge challenge to combine AI with democracy as self-rule, even in the context of a more inclusive and more tentacular body politic. Moreover, beyond classical political philosophical questions, there are also new ones: how do we live together with all these species and things, with all these non-humans? The *sympoiesis* Haraway proposes sounds attractive, but how do we do it? How do we make a better common world? What is a 'common' world once we dissolve the strong anthropocentric barrier? How does the 'politics of life agents' (Latour and Lenton 2019) pan out in practice when there are conflicts? As Latour suggests, when we find ourselves in Gaia there are not only friends but also enemies (Latour 2017, 88), for example, those who deny human agency with regard to climate change and those who refuse to share the world with others. But what about non-human enemies, for example dangerous viruses? And what are the 'sympoeietic' technologies and infrastructures we need to build it to ensure both peace/survival and flourishing? In what way can AI contribute to such technologies and constructions?

Finally, as I already said, democracy and other political artifacts, like the human beings who create it and like the ecosystems on which they depend, remain vulnerable and mortal. This has fundamental political significance. First, it gives us a reason to care for them and work on them. It gives us a reason to keep a close eye on those who try to destroy them with the help of AI and – perversely and unfortunately – do this sometimes in the name of freedom. Second, however, care does not only mean 'preserve'. It may be worth keeping the best institutions we have (democratic ones). But the mortality of our political institutions also gives us a chance to transform them. And the theories discussed in this chapter suggest that this is necessary, not only for humans and their freedom, justice, and interests, but that we should also do it for the sake of non-humans, maybe even *with* non-humans, who are not like humans but nevertheless *already* co-shaping and co-organizing the collective instead of merely being a passive part of it, and on which we *depend* for our survival, freedom, and flourishing. Politics in the 21st century is not, and should not only be, about humans; it is and should be also about non-humans, animals, climate, viruses, and indeed AI. For us Westerners, educated to think along Aristotelian definitions of the body politic and haunted by modern binaries and Hobbesian fears, these posthumanist directions may well be confusing. But if we care about freedom and other political values and if we want to deal with climate change and AI in a good way, then finding our place in, negotiating our way through, and actively co-building this new, hybrid body politic will be our task and, ultimately, the only way to move on and *become*.

References

Abbate, Cherryl E. 2016. "'Higher' and 'Lower' Political Animals: A Critical Analysis of Aristotle's Account of the Political Animal." *Journal of Animal Ethics* 6 (1): 54–66.

Arendt, Hannah. 1958. *The Human Condition*. Chicago, IL: University of Chicago Press.

Aristotle. 1984. *Politics*. Translated by B. Jowett. In *The Complete Works of Aristotle*, vol. 2, edited by Johathan Barnes, 1986–2129. Princeton, NJ: Princeton University Press.

Cochrane, Alasdair. 2012. *Animal Rights without Liberation: Applied Ethics and Human Obligations*. New York: Columbia University Press.

Coeckelbergh, Mark. 2009. "Distributive Justice and Cooperation in a World of Humans and Non-Humans: A Contractarian Argument for Drawing Non-Humans into the Sphere of Justice." *Res Publica* 15 (1): 67–84.

Coeckelbergh, Mark. 2015. *Environmental Skill: Motivation, Knowledge, and the Possibility of a Non-Romantic Environmental Ethics*. New York: Routledge.

Coeckelbergh, Mark. 2017. "Beyond "Nature": Towards More Engaged and Care-Full Ways of Relating to the Environment." In *Routledge Handbook of Environmental Anthropology*, edited by Helen Kopnina and Eleanor Shoreman-Ouimet, 105–116. Abingdon: Routledge.

Coeckelbergh, Mark. 2018. "The Art, Poetics, and Grammar of Technological Innovation as Practice, Process, and Performance." *AI & Society* 33: 501–510.

Donaldson, Sue and Will Kymlicka. 2011. *Zoopolis: A Political Theory of Animal Rights*. New York: Oxford University Press.

Garner, Robert. 2013. *A Theory of Justice for Animals: Animal Rights in a Nonideal World*. New York: Oxford University Press.

Haraway, Donna. (1991) 2000. "A Cyborg Manifesto." In *The Cybercultures Reader*, edited by David Bell and Barbara M. Kennedy, 291–324. London: Routledge.

Haraway, Donna. 2015. "Anthropocene, Capitalocene, Plantationocene, Chthulucene: Making Kin." *Environmental Humanities* 6: 159–165.

Haraway, Donna. 2016. *Staying with the Trouble: Making Kin in the Chthulucene*. Durham, NC and London: Duke University Press.

Heidegger, Martin. 1977. "The Question Concerning Technology." In *The Question Concerning Technology and Other Essays*, translated by William Lovitt, 3–35. New York: Harper & Row.

Nietzsche, Friedrich. (1886) 2003. *Beyond Good and Evil: Prelude to a Philosophy of the Future*. Translated by Reginald J. Hollingdale. London: Penguin.

Latour, Bruno. 1993. *We Have Never Been Modern*. Translated by Catherine Porter. Cambridge, MA: Harvard University Press.

Latour, Bruno. 2004. *Politics of Nature: How to Bring the Sciences into Democracy*. Translated by Catherine Porter. Cambridge, MA: Harvard University Press.

Latour, Bruno. 2017. *Facing Gaia: Eight Lectures on the New Climatic Regime*. Translated by Catherine Porter. Cambridge: Polity Press.

Latour, Bruno and Timothy M. Lenton. 2019. "Extending the Domain of Freedom, or Why Gaia Is So Hard to Understand." *Critical Inquiry* 45 (3): 659–680.
MacIntyre, Alasdair. 1999. *Dependent Rational Animals: Why Human Beings Need the Virtues*. Peru, IL: Open Court Publishing.
Regan, Tom. (1983) 2004. *The Case for Animal Rights*. Berkeley: University of California Press.
Singer, Peter. 1975. *Animal Liberation*. New York: Random House.
Vogel, Steven. 2002. "Environmental Philosophy after the End of Nature." *Environmental Ethics* 24 (1): 23–39.

Figure 7 Fence and hand.

7 The Poetic-Political Project

Challenges for Liberal-Philosophical Thinking and Practice in Times of Ecological and Technological Crisis

Summary and Conclusion: Challenges for Liberal-Philosophical Thinking in Times of Crisis

The liberal-democratic philosophical tradition is arguably one of the greatest achievements of humankind. Rooted in Enlightenment philosophy and its precursors, it has helped to create barriers against violence and oppression and has significantly contributed to the creation and maintenance of open societies. In times when, unfortunately, human freedom and human rights are *still* under threat in many parts of the world, liberalism is a welcome *pharmakon* or safety belt. However, in political history, the idea that freedom is an important value or the supreme value has always come under pressure, especially in times of social-economic crisis, war, epidemics, and so on. This has led to either the destruction of freedom or the fabrication of delicate, vulnerable, and not very stable balances and compromises that perhaps do not destroy but at least revise or compromise liberal practice and thinking. On the political left, there are socialist forms of authoritarianism but also social democracy, left-wing populism, and left politics based on identity. On the political right, there are fascist and Nazi forms of authoritarianism next to right-wing populism and right identity politics, but also more or less democratic forms of conservatism. When it comes to the stretches and compromises, the question is always, is this still liberalism or is it essentially anti-liberal? And is it still about freedom or about something else? For example, are so-called democratic forms of conservatism or right-wing populism masked forms of non-liberalism, which therefore all too easily slide into fascism and other forms of authoritarianism? Is left identity politics essentially anti-liberal given its thinking in terms of identity and groups, or is what I have called 'identity liberalism' justifiable in terms of *liberty*? And, insofar as it mainly relies on other principles such as justice and equality, is it still about liberty at all? Is there necessarily a tension between freedom and justice? What forms of intergenerational justice are acceptable? How free are people under social democratic regimes of heavy taxation? Any compromise on freedom can and has been accused of destroying freedom and not respecting the foundations of a liberal-democratic social order. Such critical questioning is not only *allowed* under liberalism but it is also necessary and desirable, since it contributes to upholding freedom as a very important, if not the most important, value in society. It turns out that liberal democracy is very vulnerable indeed and needs continuous critical questioning and perhaps also struggle, protest, and rebellion in order to survive. Moreover, freedom, in contrast to what some strands in the liberal tradition hold, is not something natural. It needs to be created and maintained. Liberal democracy is an artifact, a kind of technology or technological infrastructure. Without maintenance, it breaks down. And maintenance means continuing to discuss, contest, and protest when one believes that liberty and democracy are threatened.

Yet this book was not meant as a defense of liberty or of the liberal-philosophical tradition as such, but rather presents the results of an intellectual exercise that discusses what freedom means in the light of climate change and AI, and probes what happens if, in response to these developments, we stretch liberal thinking in a number of directions. It has addressed the problem that today, in times of climate change and new technologies such as AI, freedom, liberal democracy, and liberal-philosophical thinking are once again challenged as we face a new global crisis: an ecological and a technological one. (And in light of the Corona crisis we may add an epidemiological one.) There are two ways this crisis can pan out for liberty and liberal thinking and practice: one is the destruction of liberty and the abandonment of liberal thinking altogether, as often happened in the past and, unfortunately, is still happening today; the other one is a movement toward new balances with other values, stretching what we mean by freedom (beyond negative notions of liberty), and more generally revising the liberal-democratic model. If we want to avoid the first kind of outcome, we better work on the second one. What I have done in this book can also be called a 'third way' since it attempts to navigate between two monsters: the Scylla of authoritarianism and the Charybdis of radical libertarianism – or to use another proverb: the authoritarian devil and the deep blue libertarian sea. At some points in its journey, it also encountered other navigation problems: between liberty and other principles, and between universalist and particularist interpretations of these principles. Drawing on theory from political philosophy, the book has identified these different challenges for political liberty, and has explored Hobbesian and non-Hobbesian directions.

Let me give a brief summary of the journey. I started with the idea that in order to deal with the planetary climate crisis and prevent global disaster caused by AI, it is necessary to install a global political authority and implement measures that lead to manipulation and surveillance of people. In other words, one could offer a Hobbesian solution in order to ensure survival of humanity. However, this seems to lead to authoritarianism; can it be done in a democratic, non-authoritarian way? One response could be to nudge people, using a form of so-called 'libertarian' paternalism. But this is in turn very problematic since it does not respect freedom as autonomy and is based on a Hobbesian, pessimistic view of human nature. Another view of freedom and nature has been provided by Rousseau, whose thinking we can use to argue for democracy understood as self-rule – an idea which incurs its own problems when formulated, as Rousseau did – in terms of obedience to the general will. This discussion has led us to consider more positive conceptions of liberty, including the idea of capabilities (Sen and Nussbaum) and liberty in the context of Dewey's conception of democracy. I also raised the problem concerning the relation between freedom and ethics: classical liberalism has been too 'thin' on the good life and the common good, and connecting freedom to ethics seems important when it comes to doing something about climate change, dealing with the risks of AI, and preserving democracy. But how 'thick' can liberalism be, if it is still to be called 'liberalism'? Then we moved to discussing the problem of freedom in the context of the Anthropocene. Given the threat of climate change and an uncertain technological future with AI, we do not only need to think about the common good; it seems that we also need collective action, also on a planetary scale. But this only increases the grip humanity has on the earth, and might lead to authoritarianism at a global scale. The Hobbesian challenge resurfaces. But is freedom really the only and supreme political principle and value? What about justice, for example, climate justice or justice as fairness with regard to bias in AI? I argued that we also need other principles, if not instead then at least *next* to liberty. If we take these political principles seriously, we bounce again against, and stretch, the boundaries of liberalism, for example, when we use one of these other principles (distributive justice) to argue for redistribution of climate risk, the realization of which seems to threaten (negative) freedom, or when we consider the question regarding dependency and community: I suggested that we can interpret MacIntyre's ideas as being about positive liberty and develop them in a more relational direction. This bouncing and stretching happens even more when we consider the perhaps most 'dangerous' question of all: where shall we draw the boundary of the collective? Where is the limit of the political? What about non-humans? Should they remain excluded from the body politic? Can animals be citizens? And can things be part of the collective? What new mortal gods or what innovative political artifacts, infrastructures, and constructs do we need to expand the body politic? What technologies can help to build such new, more inclusive political worlds? I proposed that we understand the political as a poetic process, rather than a 'state' or a response to a state of 'nature'. I ended up not only with the idea of an enlargement of the political, but also with a redefinition of the political itself and a more relational view of freedom that is not necessarily opposed to (inter)dependence.

Asking and engaging with the uncomfortable questions introduced in this book is not an easy road to go down, but liberals will have to do so in order to deal with the global problems and crises we are facing as societies and as humanity. And many of us are liberals and liberal-democrats in some way or another: many of us claim to value freedom and democracy. But it is far less clear what this freedom we care about means, and what conception of freedom we need to tackle the challenges of the 21st century. If we care about the future of liberty and democracy at all, then it is important that we do not hide behind an all too easy rejection of authoritarianism (right or left) and simplistic defenses of 'freedom' in terms of negative liberty, but courageously face the intellectual and political challenges presented here and discuss potential solutions. If we fail to do so, we leave the future of our societies, civilization, and planet in the hands of those who do not hesitate to destroy the liberal-democratic order, sometimes even in the name of liberty (consider the current ongoing erosion of liberty by right populist rulers and fascist movements in the West). Then authoritarianism will happen and perhaps *is* already happening. But there are not only problems on the right of the political spectrum. I have also mentioned authoritarian forms of communism and I have pointed to the possibility and danger of a *green* form of authoritarianism, which is equally problematic in terms of freedom. In the best case, such a green form of authoritarianism or 'green Leviathan' would 'save the planet' and secure the survival of humanity (or at least some part of humanity) by creating a new green Brave New World; but it would do so at the cost of destroying human freedom and perhaps also the freedom of some non-humans.

To combine the preservation (or even better the increase) of liberty and democracy with finding solutions for global and complex problems such as climate change, AI, pandemics, biological warfare, and so on is not easy, and I hope nowhere in these pages I have suggested it is. But this book has offered some conceptual tools from political philosophy (and literature) we can use to tackle this challenge. Next to Hobbesian thinking, the concept of nudging, the idea of the invisible hand, and the idea of the Anthropocene, these tools also include some versions of positive freedom, the idea to link freedom to ethics, community, and human flourishing, thinking about global collective action, and theories that open up the political to non-humans. Consider the controversial conceptions of positive freedom as self-rule (Rousseau), capabilities (Sen and Nussbaum), participative democracy that uses improvisation (Dewey), and human flourishing through dependence and community (MacIntyre), or Latour's and Haraway's imaginative reconceptualizations of the political, which, by including things such as the hole in the ozone layer and by talking about the collective in terms of a hot compost pile or Gaia, stretch not only the body politic but also liberal thinking itself and our ideas about what politics means. In this sense, they are a kind of 'dangerous ideas'. However, given that many of the discussions in this book suggest that *only significant revisions of classical liberalism and creative ways of thinking about freedom and the body politic can tackle the problems at hand* – that is, can inspire a politics that can deal with new risks and vulnerabilities created by climate change and AI – and that, at the same time, these revisions and innovations seem the only way to avoid the perils of both authoritarianism and radical libertarianism, I conclude that, in the light of the present crises, defenders of freedom and liberalism should at least seriously consider these directions of thinking, and should stretch their liberalism before others break it.

This necessary 'stretching' includes acknowledging that the realization and maintenance of liberty and liberal democracy depends on creating the conditions for its flourishing, rather than eroding these in the name of freedom. Creating these conditions can be done by supporting capabilities, community, the common good, and the good society, next to – necessary in times of crisis – securing the very survival of the collective, which in response to global problems can only be done through collective action and cooperation at all levels, including the global level. This collective does not only consist of humans, freedom is not only about (their) freedom from constraint, and freedom is not the only value. If we continue to exclude non-humans and think that any conception that goes beyond negative freedom is a form of authoritarianism, the only freedom we will have left is a very thin, sad, and ultimately deadly kind of freedom: the freedom to defend a burning house against the fire that is consuming it, or – to revisit a metaphor I used earlier in the book – the freedom of the koala bear in a burned down forest. Ultimately, only living a *relational* and environmental understanding of freedom and realizing its conditions, including shaping the right kind of communal and societal environment, enables survival *and* flourishing of humans and humanity. If not paired to a positive one, the negative conception of freedom is not only inadequate but also dangerous. It leaves us with the choice between, on the one hand, global human and environmental disaster as a result of unrestrained negative freedom and, on the other hand, survival at the price of the total destruction of *any* kind of freedom: a Hobbesian *political automaton* that might save the planet, but leads to authoritarianism. Once we erode the social-communitarian and ecological soil on which freedom can grow, once we take away its conditions, we are left with a desert where only despotism rules: the despotism of an absolute ruler or the despotism of one's own desires (freedom of doing what you want), in other words, the Hobbesian trap. We can do better and create the conditions for a freedom, a life, a society, an environment, and a planet worth wanting.

Science and technology such as AI, then, may help us to create these conditions, but only if we acknowledge their political character and openly discuss their politics, for example, in terms of freedom and preferably in a context of Deweyan participative democracy. Technology is not only about 'technical' things. It is also about the future of human freedom, which is strongly linked to the future of non-humans, the environment, and indeed 'the planet'. Our challenge and responsibility as one of the main makers/participants, indeed poets of the body politic, is to create and employ *sympoeietic* technologies and methodologies, also in and with AI and data science. In this way, we can hope to transform the vulnerabilities of our mortal democracies into fertile humus for growing the more relational kind of freedom we need and thrive on.

This requires some more unpacking. What do I mean with *sympoietic* technologies?

The Poetic-Political Project: Politics as Poetics and a Call for *sympoeietic* AI

Time for revolution? The approach developed here implies that liberation is not just a matter of words and rebellion, but a long and largely continuous process that requires *work*. Understanding politics as *poeiesis* and as an ongoing transformative process breaks down the ancient Greek divisions between political action, work, and labor that Arendt articulated. It conveys the idea that creating the conditions for political freedom is not a matter of making oneself free from creating things and from actively relating to one's environment, as the citizens of the ancient *polis* did when they engaged in 'politics' by leaving behind their household, their women, and their slaves (politics as a kind of negative freedom), but instead demands technological action, creation, and transformation with the aim of making and remaking the collective – preferably in a radical inclusive way, that is, with the participation of all and everything (hence *sympoietic,* to borrow Haraway's term again). Politics, then, is not just a matter of declaration, agreement, and discussion, or the quasi-legalistic making of a social contract, which is all a matter of words. It also requires political technologies and infrastructures that help to make the new common world. It requires political sciences that help to gather and sustain the new collective. Marx did not materialize politics too much; he did not materialize it *enough.* Technologies and the sciences are also political. Politics may well be about communication, as Habermas argued, but 'communicative action' (Habermas 1984) is not only a matter of mutual deliberation and argumentation, that is, the performative use of words; technologies and the sciences (next to the arts) can also *communicate*, that is, they can make community and enable cooperative action. For example, action in response to climate change is made possible by discussions but also by climate science and their technologies. Politics is about words and things – the latter also have performativity and normative significance (Coeckelbergh 2017; 2019).

This reconceptualization of politics leads to a reconsidering of the role of AI in relation to the central question I ask in this book. What are the implications of this conception of politics and freedom for the role and politics of AI and data science in the light of climate change and other global crises such as a pandemic or war?

When it comes to freedom, it is tempting to only look at the direct and obvious danger of using AI in a non-democratic, authoritarian way to deal with a global crisis such as climate change. This discussion is the home of the Hobbesian Leviathan metaphor. I evoked this way of framing the problem in my story in the beginning of this book. Unfortunately, the Hobbesian solution is also partly reality: as I write, all over the world and even in Europe, a crisis situation (the Coronavirus pandemic) is used as an excuse for restrictions to the liberty of citizens and in some cases even the disempowerment of democratic institutions. Consider the authoritarian response of China or the dictatorial twist in Hungary. China is already using AI and data science for extensive surveillance. One can imagine similar developments under a green kind of authoritarianism. Such developments concern, among other things, the impact of AI on negative freedom, and the effects are often intended. Concerns about the infringement of these liberties are justified. We need to ask the question, Do people, corporations, and governments (try to) restrict our liberty by using AI, and under what conditions (if at all) is this justified?

However, if we embrace the alternative, positive, relational, and 'poetic' approach to politics and freedom I proposed, and take on board the insight from the philosophy of technology that technologies also have a non-instrumental character, we must also pay attention to different and more subtle, non-intended political effects of AI. Once we consider the question concerning freedom as also including the problem of its conditions and as being linked to the process of gathering and maintaining the collective, we are no longer stuck in a merely Hobbesian problematic; we can also look at the politics of AI and its relation to freedom in terms of its *sympoietic* effects. We can then ask a new question: Does this technology and the related science(s) gather us together, connect us with other entities, and help to create the conditions of our flourishing – thus contributing to positive and relational freedom and to the poetic-political project of building a common world?

In light of the climate crisis and the possibilities offered by AI, the answer we can presently give to this question is twofold. On the one hand, there are reasons to answer positively and to be optimistic: one could point to the new *sympoietic* and relational opportunities created by AI and data science for dealing with the climate crisis. First, with regard to the climate problem, these technologies and sciences can show us that 'we are all in this together', that is, we are all affected by and dependent on these changes and this crisis. The 'we' here means humans, but also non-humans. AI and data science can reveal our strong interconnectedness: it can show that humans and non-humans are both interdependent and dependent on the same ecosystems and living on the same planet. Understood as *political* technologies and sciences (with politics defined in a relational way), AI and data science can thus promote global-ecological thinking. Second, with regard to the solution, AI and data science can help us with the global coordination and communication actions needed to deal with the problem. Together with climate science, AI and data science can help us to gather around the problem (climate change) *and* around a solution. They can help us to build a new political collective around the climate crisis. They can function as tools in the hands of the human poets that make, and participate in, a new world. This is a *sympoietic* project since we can only do this together: together with other humans and with the help of technologies such as AI.

On the other hand, there are also reasons to be concerned, especially if we look at the subtler effects, the non-intended consequences. Usually these problems are defined as 'ethical' problems: AI has consequences for privacy, for the manipulation of people, for responsibility, for the lack of transparency of decisions, for victims of bias, and so on (Boddington 2017; Dignum 2019; Coeckelbergh 2020). Often the focus is on effects on individuals. But these ethical discussions also show that there is always a societal aspect to these problems. Consider, for example, how the question regarding bias leads us not only to consider the harm done to specific individuals but also to consider questions concerning justice and equality at the societal level. In light of the approach developed in this book, one could say that these ethical problems also have a *political* dimension, in the sense that AI can help us gather the collective but can also *negatively* impact the gathering and building of the collective and hinder rather than promote human and non-human flourishing. It can produce more fragmentation, create less fairness, less transparency, etc., threaten the flourishing of humans, and also negatively impact non-human lives such as the lives of non-human animals. With regard to climate change, for example, AI used for oil extraction can directly or indirectly disrupt human and non-human communities by means of its immediate local environmental effects, but also by having effects on climate change and by increasing the risk of violence and war. This hinders the gathering of the political collective at the local and global level and destroys rather than builds common worlds.

Our response to these problems (as humans and as humanity) should not only be more discussion but doing more *work*. The challenge for us as poets of the political, who in the present situation carry a big share of the responsibility for the planet and its climate, is not only to discuss the problems but also to identify and enlarge the political-poetic opportunities. The adequate response to the problems posed by AI in the light of global crises such as climate change is *not* to be against AI and data science, but rather to recognize their political significance and potential, next to reducing the mentioned risks and addressing the ethical and political problems. Guided by a non-Hobbesian, *sympoietic* imagination and implementing this relational approach in practices of technology development and use, let us not only worry about authoritarian uses of AI and the survival of humans and humankind, but also integrate these technologies and sciences in the larger positive and relational political-poetic project of liberation and democratization by having them contribute to human and non-human flourishing, the making of a more inclusive collective, interdependent self-rule, and the building of new common worlds. If this project fails, we will not only have destroyed the conditions for freedom and democracy, understood in a positive and relational way; ultimately, we will also have killed the political itself.

References

Boddington, Paula. 2017. *Towards a Code of Ethics for Artificial Intelligence*. Cham: Springer.

Coeckelbergh, Mark. 2017. *Using Words and Things: Language and Philosophy of Technology*. New York: Routledge.

Coeckelbergh, Mark. 2019. *Moved by Machines: Performance Metaphors and Philosophy of Technology*. New York: Routledge.

Coeckelbergh, Mark. 2020. *AI Ethics*. Cambridge, MA: MIT Press.

Dignum, Virginia. 2019. *Responsible Artificial Intelligence*. Cham: Springer.

Habermas, Jürgen. (1981) 1984. *The Theory of Communicative Action*, vol. 1. Translated by Thomas McCarthy. Boston, MA: Beacon.

Index

For Product Safety Concerns and Information please contact our EU representative GPSR@taylorandfrancis.com
Taylor & Francis Verlag GmbH, Kaufingerstraße 24, 80331 München, Germany

www.ingramcontent.com/pod-product-compliance
Lightning Source LLC
Chambersburg PA
CBHW060317220326
41598CB00027B/4352

* 9 7 8 0 3 6 7 7 4 7 7 9 4 *